THE OCEAN'S
WHISTLEBLOWER

THE OCEAN'S
WHISTLEBLOWER

The Remarkable Life
and Work of *Daniel Pauly*

DAVID GRÉMILLET

Translated by
GEORGIA LYON FROMAN

DAVID SUZUKI INSTITUTE

GREYSTONE BOOKS
Vancouver/Berkeley

First published in English by Greystone Books in 2021
Originally published in French as *Daniel Pauly: Un océan de combats*,
copyright © 2019 by Éditions Wildproject
English translation copyright © 2021 by Georgia Lyon Froman

21 22 23 24 25   5 4 3 2 1

Greystone Books Ltd.
greystonebooks.com

David Suzuki Institute
davidsuzukiinstitute.org

Cataloguing data available from Library and Archives Canada
ISBN 978-1-77164-754-0 (cloth)
ISBN 978-1-77164-755-7 (epub)

Copy editing by Paula Ayer
Proofreading by Jennifer Stewart
Jacket design by Fiona Siu and Jessica Sullivan
Text design by Fiona Siu
Jacket photograph by Sherman Lai

Printed and bound in Canada on FSC® certified paper at Friesens. The FSC® label means that materials used for the product have been responsibly sourced.

Greystone Books gratefully acknowledges the Musqueam, Squamish, and Tsleil-Waututh peoples on whose land our office is located.

Greystone Books thanks the Canada Council for the Arts, the British Columbia Arts Council, the Province of British Columbia through the Book Publishing Tax Credit, and the Government of Canada for supporting our publishing activities.

For Yvonne and Jacques, who like biographies.

TABLE OF CONTENTS

I. ORIGINS

A Swiss Childhood   1
Youth in Germany   19
The Search for an American Father   32

II. CONSTRUCTIONS

From Oceanography to Fisheries Biology   42
Daniel's First African Experience   55
Development Aid in Indonesia   65
Birth of a Career in the Philippines   83
San Miguel Bay and the Social Dimension of Fisheries   89
Tropical Statistics   98
A Pacific Heroine   105
A Man of Letters   112
Fish Stories in Peru   119
Nature in a Box   135
For All the Fish in the World   149

Photo Album   161

III. ON THE WORLD STAGE

The Big Leagues   186
Fishing Down Marine Food Webs   206

The Sea Around Us   225
Chinese Fisheries and Charles Darwin   240
Uncertain Glory   247
Reconstructions   259
Africa Forever   263
French Allies   284
First Loves, Final Battles   298

Epilogue   *308*

APPENDICES

Important Dates in the Life of Daniel Pauly   316
List of Abbreviations and Organizations   318

Notes   *322*
Selected Bibliography   *334*
Acknowledgments   *338*
Photo Credits   *340*
Index   *341*

NOTE FROM THE TRANSLATOR

All notes included in this book are the work of the author, except those signed "TN," for "Translator's Note." Additionally, for the reader's convenience, when relevant, metric measurements in the text are followed by their equivalent in customary units rounded to the nearest whole number. Monetary amounts are in US dollars unless otherwise specified.

I

ORIGINS

A SWISS CHILDHOOD

RENÉE HAD A limp, all because of the bombs. The ones she manufactured in Berlin, then in Frankfurt, and the ones the Allies dropped on her factory one February day in 1944. That year, hundreds of bombers came to pound the German industrial basins, the churches, and the population. Some of the victims were from among the two million French men and women who worked in Germany during the war, willingly or otherwise. Renée crossed the border voluntarily, along with forty thousand of her female compatriots.* French history, always careful to emphasize resistance against the enemy, hardly mentions these women, but German scholars have revealed their existence. They rarely went out of conviction, but rather to flee poverty, the law, or their families. The German occupiers did everything in their power to "reduce unemployment" in the French regions and recruit cheap labor for their factories.

Renée left the French city of Nancy with the promise of training and better work conditions—and a strong desire to get away from her father, Henri Clément. Henri was a brute, a big, straitlaced policeman who had fought at Verdun and who worshipped Marshal Pétain. Shortly before the First World War, he married Marie Kaltenbach, whose Jewish family in Lorraine wanted nothing to do with him. The couple settled in the Champagne region and had nine children. Renée

---

* The *Service du travail obligatoire* (Compulsory Work Service, or STO) only applied to men. French women who worked in German factories were therefore necessarily volunteers.

was the fourth, born in 1923. She had her father's long face and her mother's steely gaze and sharp mind. Everyone agreed that Renée should continue her studies, and her schoolmistress supported her as far as she could. But her father wouldn't hear of it: "She'll work like the rest of them," he declared. And so, at the age of fifteen, Renée began her first job as a domestic in Nancy.

When that American bomb came down on her factory in Germany, Renée was hit in the leg. The woman next to her was killed. In the hospital, Renée risked gangrene and amputation, remaining bedridden for six months. During her convalescence, she was given work mending Wehrmacht uniforms. Because of fabric shortages, by the last few weeks of the war, they were made of crepe. As American troops surrounded Frankfurt, she witnessed the nurses discreetly remove their Nazi armbands. Renée survived the fall of the thousand-year Reich only a little worse for wear and returned, limping slightly, to France. For the rest of her life, she would receive a small disability pension from the German state.

She was back at her parents' house in Avize by the summer of 1945, when the American military convoys came rolling through on their way east. Ninety miles to the south, in the Yonne, my mother, then fourteen years old, also saw them pass by. She remembers seeing the big Black men with their wide smiles riding atop the trucks. Blacks and whites did not mix in the US Army—the dark-skinned troops formed separate regiments commanded by white officers. These segregated units were generally unarmed and restricted to logistic support. Yet African American soldiers had already sought out combat roles and proved themselves in battle: during the First World War, the Harlem Hellfighters of the 369th Infantry Regiment, aided by the French, demonstrated their fighting prowess at the Battle of the Somme. As a result, the first foreigner to receive the French Croix de Guerre was an African American soldier.* It was not until much later, in 1948, at the beginning of the Korean War, that President Truman officially

---

* Henry Lincoln Johnson.

abolished segregation in the American armed forces.* In the mean-time, many a military ambition went unfulfilled, though some African American soldiers pursued their dreams abroad. The most famous of these was James L. H. Peck: when he was not allowed to become a pilot at home, he went to fight in Spain, beginning in 1936. At the controls of a Russian-made Polikarpov fighter, he brought down a dozen fas-cist planes—including two from the Condor Legion, which had been responsible for the attack on Guernica—becoming the first Black fighter pilot in history. In 1941, a change in American law allowed a unit of African American pilots to begin training in Tuskegee, Alabama. After a half-hour test flight during an official visit to Tuskegee, First Lady Eleanor Roosevelt declared, "Well, you can fly, alright!" The group would go down in history as the Tuskegee Airmen, participating in the Italian campaign aboard their famous "Red Tail" Mustang fighters.

Winston dreamed of joining them. He was born in 1919 in Little Rock, Arkansas, with the full name Winston R. McLemore. Though he wanted to be a pilot, Winston was a little too impulsive, a little too bad-tempered—and most significantly, he drank like a fish. He never joined the ranks of the elite Black airmen, but instead drove trucks during World War II before reenlisting to serve in Korea and Guam.

In 1945, the American soldier was passing through Avize. Winston met Renée, and the details of their brief affair are theirs alone. Renée was open about their relationship and the ones that followed. This was also a way to defy her father, and she did not hide the fact that the child she was expecting would be mixed race. Winston left for Germany, then returned to the United States. Renée's father kicked her out of his house. She went to Paris alone and spent a few months working at the Saint-Maurice psychiatric hospital. Then, on May 2, 1946, north of the capital, a little boy was born. She named him Daniel Marc Roger. The life of a single mother, doubly stigmatized by everyday racism, was brutally difficult. Renée worked a factory job and entrusted Daniel,

---

* Executive Order 9981: "There shall be equality of treatment and opportunity for all persons in the armed forces without regard to race, color, religion, or national origin."

more or less successfully, to the care of various babysitters. The Parisian winter, insufficient ration cards, smoke from the factories, a baby passed from one pair of hands to the next—Daniel was constantly ill. A single photo survives from this period of his life: a wide-eyed toddler with brown hair opens his mouth in delight as he clings to the bar of an antique high chair. The family of two only came together in the evenings and on Sundays.

One summer day in 1948, Renée and Daniel spent a few hours on the banks of the Seine at the Quai de Bercy. A little blond girl played with Daniel while the grown-ups waded in the water. The girl was from Romandy, the French-speaking part of Switzerland, and she had come to Paris with her parents and two older brothers for a family vacation. Much later, Daniel would call them "my Thénardiers,"* but I suggest calling them the G---s. Madame G chatted with Renée, presenting herself as a respectable person from La Chaux-de-Fonds. She had come to Paris with her husband (a man of few words) to visit her sister, who lived in Belleville and owned a café with her Algerian husband where immigrants drank tea and played dominos. The G---s thought Daniel was adorable, and Renée opened up about how hard life had been for her, how she worried constantly about her sick baby boy and was struggling to raise him alone.

The exact terms of the agreement that followed are unknown. During her long life, Renée would give slightly different versions of it. What is certain is that the G---s, who had recently lost a son Daniel's age to meningitis, offered to take the little boy back to Switzerland with them for a few months to give his exhausted mother a break while Daniel recovered his health. Renée accepted the offer, and the G---s went back to La Chaux-de-Fonds with Daniel by steam train, stopping only once, at Besançon.

---

* Monsieur and Madame Thénardier are the main antagonists of Victor Hugo's 1862 novel *Les Misérables*. They are portrayed as a couple of petty criminals with brutal and abusive tendencies. (TN)

WITH A MEAN annual temperature of 43 degrees Fahrenheit, La Chaux-de-Fonds is an austere town spread along the floor of a Swiss valley 3,280 feet above sea level. With its tidy grid-like layout, it reminds me of a military base, its large buildings sporting endless rows of windows, like so many eyes watching the town's watchmakers toil within. Largely working class and politically communist, the town's residents have long fought to secure labor rights, as well as many progressive social and cultural policies that can still be seen today.

I arrive there on a sunny afternoon in May of 2016. The train from Geneva runs along the edge of a lake until it reaches Neuchâtel, then climbs straight up the mountain to La Chaux-de-Fonds, a thirty-minute ride that turns out to be very painful for my sinuses. "Neuchâtel is only twelve miles from La Chaux-de-Fonds, but the trip was always an expedition," Daniel remembers. "Now, there's a highway tunnel, but at the time, the only way out was through the pass of Vue des Alpes. As soon as there was a lot of snow, we would be cut off from the world, sometimes for several weeks."

I end up in a shabby hotel near the train station—the room is dimly lit and the bed is round, like in a brothel. The neighbors are listening to bad rap music on high volume, and as I walk down the hall, I hear a woman shout from her room: "I'm comin' with five Albanians! I'll kick your ass!"

Though its oversized, austere main street reminds me of a Soviet-style city, La Chaux-de-Fonds is actually a small town. In fact, the countryside is so close that a slight odor of cow manure wafts constantly through the streets. The European metropolises that tend to attract immigrants are a long way off, but the population here is quite diverse. Already, in the train, I was seated across from an African man, likely an Eritrean refugee. In town, school is letting out and a whole bevy of Black schoolchildren pours into the street. Later that evening, I see a jovial Congolese family leaving the municipal theater.

Thanks to some of Daniel's old Swiss friends, I was able to track down the G---s' children. The next day, I meet their daughter, who is now seventy-eight years old. "When we came back to La

Chaux-de-Fonds at the end of the summer of '48," she recalls, "every-one thought my mother had had a baby with an African in Paris. I was very happy, though. I took him everywhere in his pram—he was the town's main attraction, the only Black child we had ever seen in La Chaux-de-Fonds." The G---s lived on Rue du Puits, on the northern fringes of the town center, across from a gas plant. Their apartment, in a prisonlike five-story building, was government-subsidized, and they were rag-and-bone men by trade. Daniel was another mouth to feed in this state of near poverty, and it seems surprising the family would have agreed to take on the extra burden so easily. In France, as in Swit-zerland, I am assured that the G---s never took a penny for Daniel's care. Some even tell me they were motivated by a feeling of "solidarity among the down-and-out." But the G---s soon cut ties with Renée—they were supposed to keep Daniel for three months, and initially, they sent news about the boy. When Renée asked to have him back, though, they refused. But she didn't give up easily, sending letter after letter to the G---s.

In 1949, Renée married Louis Pauly, whom she had met at the Compagnie française des métaux, a French metallurgy company, in Paris. Louis had been raised by social services—after growing up in various orphanages, he had generous ideas about family ties, and he adopted Daniel without even meeting him. "I could have had you with an Antillean woman," he would tell Daniel much later. And so, Daniel Clément became Daniel Pauly. Renée informed the G---s and asked again to see her son. By way of answer, they threatened to report her for child abandonment. The little people of France have never trusted the police or the state, so Renée let it go, telling herself that Daniel was better off in Switzerland. She went on to raise seven children with Louis Pauly in public housing in Maisons-Alfort, where her other children would see her cry every year on the second of May without knowing why.

Sometimes, however, Daniel wasn't far from his real family. When Madame G visited her sister in Paris and left the rest of the brood in Switzerland, she would bring Daniel along—without alerting the

Paulys. Daniel remembers that he came home from these trips with fleas, and Madame G with crabs (there were few secrets in the G--- family). Growing up in La Chaux-de-Fonds, Daniel never received any of the numerous letters and packages that his mother sent for him, though no one tried to hide his Franco-American origins either. "They always let me know that I was different. They told me over and over again that my mother had rejected me, that I should be grateful, that without them, I would have died of hunger. Little by little, their grief over losing their youngest son turned into greed, and I became their servant. As soon as I was old enough to work, I was seen as a resource."

DANIEL DOESN'T REMEMBER much about his early childhood, but some things do stick out. One day, he was in a shop trying to exchange some bottles for the deposit when the cashier asked him why he wasn't in school. Another day, the school nurses came to give checkups and asked all the little boys to strip down to their underwear. Daniel was embarrassed—he didn't have any. Later, something similar happened during vaccinations when a nurse cleaned Daniel's arm with a cotton swab. The cotton came away all black, and his skin changed color. At home, his foster mother yelled at him: "You piece of shit kid, I'll teach you!" And after the words came the blows, and after the blows, the closet or the basement.

When I ask her about it, the G---s' daughter denies everything: "Our *moutre** never hit anyone." Tightening her grip on her walker, she tells me, "No, Daniel didn't get any *schlagées*.† No, we didn't lock him in the basement, he was nicely dressed and coddled... No, it's not true that he didn't have a coat in the winter... We liked Daniel—he was a little brainiac, I called him my little lawyer. Look how sweet he was." And there, in her cabinet with the other family photos, she points out a frame with four photographs of Daniel as a child, looking serious

---

* *Moutre*, from the German, is a colloquial term for "mother" used in French-speaking Switzerland.

† *Schlagée*, another colloquial term used in French-speaking Switzerland, is from the German *Schlagen*, meaning "beatings."

enough to make a person cry. You would think they had been there for decades, but I know that the G---s' daughter has taken them out specifically for our interview: on the phone beforehand, she told me, "I'll look . . . I must have some photos of Daniel somewhere." Then the eldest brother arrives. At eighty-four years old, he has survived three bouts with cancer. Maintaining a united front, the two of them tell me a pretty story: "We got along well with Daniel, we never understood why he left without saying goodbye, we'd like to have his address so we can write to him." At last, it's their youngest brother, interviewed in a comfortable café, who confirms Daniel's version of events: "May she rest in peace, but my mother was a horrible person, and she beat Daniel." He chooses to break the silence because he took a few good *schlagées* himself back in the day. We discuss Daniel's childhood at length— the cat is out of the bag. More recently, one of the G---s' relatives also came forward to testify to the fact that Daniel had been mistreated, remembering a little dark-skinned Parisian boy who spent a great deal of time on the sidewalk out in front of their building.

> The interesting thing is that almost all the children who did survive very quickly elaborated "theories of life" that combined dreams and intellectualization. Almost all resilient children have to answer two questions. Asking "Why do I have to suffer so much?" encourages them to intellectualize. "How am I going to manage to be happy despite it all?" is an invitation to dream. If this inner determinant of resilience can find a helping hand, the prognosis for these children is not unfavorable.*

Primary school would be Daniel's first refuge. After a great deal of searching, I manage to find the names of his first three schoolteachers: Mademoiselle Lorette Lüdi (1953–54), Madame Pierre Pantillon (1954–55), and Madame Eva Grandjean (1955–56). These are the women who taught Daniel how to read and write. The Freinet Movement, an

---

* Boris Cyrulnik, *Resilience: How Your Inner Strength Can Set You Free From the Past* (New York: Penguin, 2009).

approach that encourages children to express themselves freely, was all the rage among Swiss teachers at the time. Though I can't prove it, I believe that Daniel benefited from this tolerant teaching style. In particular, his teachers never tried to change his left-handedness. "I learned to write on a slate, and I remember that, as I went along, I would erase everything I'd just written—my palm was covered in chalk. Later on," he continues, "I enjoyed the rapid mental calculation contests, which are still helpful to me today." During Daniel's first year of school, his health remained fragile: his records show fifty-two absences for illness and ten for convalescence. Things improved during his second year, but the year after that, he racked up thirty-eight absences for "helping parents."

In fact, Daniel regularly joined Monsieur G on his daily rounds in his red van to pick up all sorts of recycled materials. They also put him to work cleaning out the apartments of people who died without succession. The G---s stored then resold everything: furniture, rugs, clothes, dishes, books...

Daniel washed whole vanloads of plates, but he also salvaged several cubic meters' worth of books. Once he'd gotten started, Daniel read absolutely everything, "from the highest to the lowest literature." As soon as he got home, someone would shout at him to get to work, but he read in secret. "*Turli!** Lazy boy!" became the war cry of the *Moutre*, who was after him constantly. "If there's one habit that I kept from that time," he says, "it's that I always feel guilty when I'm not working." The books and the wisdom they contained became Daniel's refuge. He was already the know-it-all of the family and had no other hobbies besides sledding now and then, or picking mushrooms with Monsieur G, "a good man who was dominated by a terrible shrew."

Daniel made it from primary to middle school by forging the G---s' signatures on the registration form. Thanks to pressure from his teachers, his foster family gave in and let him continue his education, but he would have all sorts of odd jobs over the years. After school, he worked

---

* In bad French, short for "*Tu relis!*": "You're reading again!"

in a pastry shop, then in a secondhand clothing store. In all, he brought home a hundred Swiss francs a month to the G---s. Around that time, social services moved the family to Rue Fritz-Courvoisier, then to Rue du Collège (just a stone's throw from their old home on Rue du Puits). On Rue Fritz-Courvoisier, they lived in an enormous building that had been a boarding house for young girls. When they moved in, it was still being used to store furniture and other equipment for the mayor's office. Over the years, the G---s sold everything in the building, even the bronze lettering that spelled the name of the establishment across the front. Aside from that, they weren't big-time criminals, though they did mix with all sorts of petty crooks, burglars, pimps, and prostitutes of both sexes. On Rue du Collège, where they moved after that, there was a lot of drinking. "It was the main drug available at the time. I saw all of it, but the G---s made sure I understood that I wasn't one of them. They put a wall up between me and their shadier activities. They never let me be an accomplice," Daniel remembers. "And it was a good thing, too, because I probably would have turned out badly. But as a result, I didn't get into that much trouble after that, and I never smoked or touched a drop of alcohol. All those things that young people have to learn for themselves, I never had to learn."

Daniel could only stand so much, and when he was ten years old, he tried to run away. With two francs in his pocket, he set out for France to find his mother, but his escape attempt stopped short on the banks of the Doubs River, which he was too afraid to cross. That same night, he was back on Rue du Collège. Unable to reach France at that age, he sometimes ran away to a friend's house instead. "We would get to his house, and his mother would give us each a glass of milk and let us play. I couldn't believe my eyes because when *I* would go home, they would yell at me and tell me to get to work."

But if there is a wicked witch in this story, there is also a fairy god-mother, and Daniel's was every bit as pretty as the ones in the movies. She makes her first appearance in his earliest childhood memories. He was three or four years old, running around a fountain downtown across from the Migros supermarket. Marguerite Rognon, who was twenty years old at the time, worked at Migros, and with her, Daniel began living a double life. "At the G---s', I was dirty, dressed like something the cat dragged in, not well cared for. But from time to time, I would go stay with Marguerite's family. Her father was a foreman in a watch factory and her mother was a housewife. I would take a bath, get dressed in clean clothes. I learned table manners, all the things that make a child into a civilized person. Just knowing that another world existed—that was enough to keep me from being broken."

Marguerite also gave him his first dictionary, an illustrated *Nouveau Petit Larousse* from 1952, which Daniel still has. He would read it obsessively, then drive his teachers crazy by articulating some unusually decisive opinions on semantics during class. When he was eight and nine years old, Marguerite took him on vacation with her to Tarragona, near Barcelona, two summers in a row. By then, she had married Mateo Roqué, a watch salesman from there. Daniel made friends with one of the family's little nephews, and they had a lot of fun together in Catalonia. "They were so kind to me. I was treated the way all children should be treated," Daniel recalls. But at the end of the first trip, he turned difficult, crying and fighting, angry without knowing why.

In 2008, when he received the Ramón Margalef Prize in Ecology, Daniel saw Catalonia again. In the speech he gave at the award ceremony in Barcelona, he talked about visiting the region as a child. "And I started crying in front of everyone—it was very embarrassing. I thought about that eight-year-old child who didn't want to go back to Switzerland, and I felt so sorry for him." But when Daniel was twelve or thirteen years old, things started to look up. He had finally gotten big enough to fight off the *Moutre* when she tried to slap him around, and he freed himself from the constant work, finally carving out some time for fun. With a few friends, he joined one of the many clubs in town where young people could make model airplanes using little two-stroke engines that choked and sputtered. Around that same time, on the other side of the world where he was stationed in Guam, Winston McLemore, for want of opportunities to become a fighter pilot, also enjoyed making model airplanes.

SUMMER, 1960. The city of Lausanne would have liked to organize
the Olympic Games, but they ended up taking place in Rome during
a heat wave of historical proportions amid extreme tension between
the United States and the USSR. The American delegation got off to
a rocky start, the Soviets trouncing them thoroughly during the first
week. But African American athletes saved the day: Cassius Clay, an
eighteen-year-old boxer, won the heavyweight gold medal that
launched his legendary career. Above all, it was one Wilma Rudolph
who made headlines. Born twenty years earlier in a Black neighborhood in Clarksville, Tennessee, the twentieth of twenty-two children,
she survived double pneumonia, scarlet fever, and a bout with polio
that left her in a leg brace. Going directly from physical therapy to
intensive sports training, Miss Rudolph made a series of fabulous wins
in Rome. Ultimately, she went home with three gold medals, one for
the 100 meters, one for the 200 meters, and one for the 4 × 100-meter
relay. A black-and-white movie immortalizes the historic 100-meter
race where she finished three meters ahead of all the other competi-
tors: she dashes past the finish line, then slows down, looking straight
ahead, then at the ground. Finally, she turns around and heads for
the locker room. No expression whatsoever, no wave to the crowd—
Wilma looks as though she's just finished a training session. Her hair is
cut short, her physique incredible, with legs that make up two-thirds of
her height. Back in the United States, she would force her hometown's
segregationist mayor to organize a celebration where all of the city's
residents would be invited, regardless of the color of their skin.

In La Chaux-de-Fonds, Daniel heard talk of the Olympic Games in
Rome, but at the end of the summer of 1960, his mind was elsewhere.
It was time for the grape harvest, and he had decided to participate.
The hills of Neuchâtel produce an excellent Chasselas and the harvest
is an important social event: boys from good families sign up months
in advance. "It was the kind of activity that was completely out of
reach for a guy like me, and I was tired of always being left out," Daniel recalls. Using a visit to one of the G---s' sons in Neuchâtel as an
excuse, Daniel showed up at the gate of one of the big wine-growing

estates. The main house was empty; everyone was out working in the vineyards except for an elderly woman who greeted him bluntly: "It's too late now, you should've asked six months ago." But suddenly, her attitude seemed to change: "Say, it's funny, you look a lot like Wilma Rudolph. Why don't you run down to the vineyard and see if they have something for you?" And so Daniel has one of the pioneers of the Black Consciousness Movement in the United States to thank for being able to participate in the Swiss grape harvest of 1960. He earned a hundred francs and bought himself his first record player, along with some 45s, including Roy Hawkins's "The Thrill Is Gone" and Ray Charles's rendition of "Hit the Road Jack."

Despite a chaotic early education, Daniel was accepted into La Chaux-de-Fonds's business high school in 1960. The establishment was housed in a pink sandstone structure reminiscent of a pastry confection, located high up on the west side of town, about fifteen minutes and 135 imposing steps up from Rue du Collège. It was a prestigious technical high school, set up to train future bank directors and other notables.

"I'll never understand how a literary type like Daniel managed to land in that school instead of a normal high school," his classmate Suzy remembers with astonishment. At the time, Suzy was Mademoiselle Mauerhofer, and her parents owned one of the largest bakeries in town. Suzy was always first in her class, the blondest, the most expressive, the most authentic, the most headstrong of all the girls from good families. "She was perfect," Daniel remembers.

A half-century later, Suzy's eyes still sparkle when she talks about the young Pauly and their school years between 1960 and 1962. She is married to Jean-François Blaser, another classmate, whom everyone calls Jef, and they split their retirement between Switzerland and the middle of nowhere in the Ardèche region of France. Jef had a serious heart attack in 2015, and they decided to renew ties with Daniel. Fortunately, they contacted him soon after I began my research for this biography, in which they became active participants. They are the ones who found the G--- children when I was about to give up all hope of interviewing them. One of their main sources of information was the online archives of the *Feuille d'avis de Neuchâtel*, the first francophone newspaper in Switzerland, established in 1738. Even more surprisingly, they were able to track down one of their old teachers from the business school, Jacques Barbier, who now lives in Morocco. Thanks to the Blasers, I was up to my ears in unusually sprightly septua- and octogenarians.

I START UP my old bucket of bolts and drive several hours down the twisting country roads between Montpellier and Lagorce, where the Blasers have been restoring an eagle's nest above the gorges of Ardèche for the last twenty-five years. They give me a fascinating account of Daniel Pauly as a high school student.

"He was svelte, handsome, with an explosive laugh. Like a lot of teenagers, his legs were so long compared to the rest of his body that he looked like he was walking on stilts," Suzy says. She also tells me about his hands—"As big as a delivery doctor's, or a pianist's!"—and his ragged clothes. "I can still remember that gray sweater he wore summer and winter, with sleeves that were too short, the stitching a complete mess, and a big hole in the front. His pants were just as bad." The Blasers ruminate for a while and conclude: "We never did see him in a coat, not even in the middle of winter when there was so much snow on the side of the road that it covered the cars."

Jacques Barbier backs them up: "Suzy and Jef were from very bourgeois families compared to Daniel, who literally came from the very bottom, from a very unsophisticated family. He certainly wasn't well treated, he was punished, yelled at, beaten some."

"La Chaux-de-Fonds must have been like a prison for him," Suzy concludes, "but he wouldn't stop fighting."

In class, Suzy sat just behind Daniel, which turned out to be very entertaining, at least for her. "He was a troublemaker, a joker, a real clown who didn't do any of his work. He spent most of his time disrupting class and making us laugh. That's a gifted student for you; he must have been bored to death." The beautiful Suzy suddenly drops her reserve: "Daniel has always bothered idiots, and we had some teachers who were real jerks, who criticized his clothes or the color of his skin. Daniel talked back, of course. He did it without being aggressive, but that didn't help anything. It was incredible how he drove the teachers crazy."

"He was more intelligent than most of his teachers, which is always problematic," Jacques Barbier adds politely. "Daniel knew how to mock people, but with such subtlety and sensitivity that they couldn't get mad. Only a math teacher would lose his head over something like that!" As a student, Pauly was vexing—he questioned everything, even things that didn't affect him. When a new circular came around declaring that it was forbidden to smoke within the school walls, Daniel—who didn't smoke then and never would—promptly took himself to the principal's office to ask "on behalf of his classmates" if the imposing steps leading up to the school building were included in the nonsmoking area.

During his years at the business school, Daniel showed no interest in scientific subjects, nor in anything having to do with nature. In his language classes, he wasn't much better—the German teacher, not exactly a thrilling character, was promptly nicknamed *Schneckelette* ("Little Slug"), and Daniel remained totally immune to declensions. Because it was a technical high school, there were also classes in industrial drawing. Daniel, despite being very handy, spent most of his time spreading the contents of the inkpots on his classmates' papers. The same chaos reigned in gym class, where he often hijacked the various exercises, using the climbing ropes to board imaginary vessels. Daniel's only safe harbor turned out to be Jacques Barbier's French and history classes. "Daniel owes everything to that teacher," Suzy concludes. "The others were very tough, they got on our last nerve," insists Jef, who also struggled during those years.

In French class, Daniel finally felt like a person. "He had a very rich vocabulary, and he didn't mind writing essays, he just wrote and wrote," recalls Barbier, who was a young professor at the time, only "ten years and couple of lessons" ahead of his students. He continued teaching for only a few years before starting a consulting firm in urban development that led him to travel the world. When he lost his wife at sixty, he said to himself, "Jesus Christ! I don't want to spend the next twenty years walking the dog and waiting for it all to be over." He threw death and retirement out the window and moved his office to Morocco, where he started a new life. At eighty-four years old, he is the father of two young teenagers.

Watching his own children, he remembers Daniel, "a gangly young guy, a smart cookie, extremely likable." A student from the worst neighborhood in town, he didn't have a lot of allies among the other teachers or students. "I have to take care of this kid, he's just too smart to get into trouble," Jacques Barbier told himself. So he went down to Rue du Collège, where Monsieur G greeted him from afar with a "Hey Jacquot! What do you want? We're working here!" Barbier realized then that Daniel really was "a cuckoo that had fallen into the wrong nest."

"He never does a damn thing!" the G---s told him grouchily. "He comes home, does his homework, then he reads, and he always wants to talk, but we don't have time for that. If we give him work to do, he does it with one eye on a book. All he's gotta do is stop school and work with us." Barbier negotiated hard and won: for the next year, when school let out, Daniel would go to the Barbiers' house to do his homework and eat dinner, then back to Rue du Collège to sort copper scraps and old light bulbs.

It was a good, if short, time for the Barbiers, too. "We laughed a lot. He and my wife discussed current events. With all that, his grades went up, and he became a little more manageable." Daniel was overjoyed—"At your house, we talk at the dinner table!"

After a few weeks, he offered the Barbiers a gift: "Here, I made you a machine that turns itself off!" The young couple discovered a mechanism built into an old cigar box with a battery, a little model airplane motor, and a tangle of wires. Daniel showed them how it worked: when he pushed a button on the outside, the box opened and the motor started up, then a little lever lifted a weight which fell onto a second button, turning the machine off again. Jacques Barbier remembers every detail of that pataphysical device, and it still makes him laugh.

But the teacher was still worried about his student's future. "In La Chaux-de-Fonds, he could have become a mailman at best. You could also have imagined him getting into drugs or petty crime. But Daniel had a real distaste for everything that was going on around him; he knew he wasn't meant for this."

With the help of the school principal, Monsieur Jeanneret, Jacques Barbier looked into a long list of possibilities and considered sending Daniel to stay with a German nonprofit that cared for children of the war. But the Barbiers left La Chaux-de-Fonds for Neuchâtel before the plan could come to fruition, and Daniel's world fell apart: he lost his only ally at the high school and then, in the spring of 1962, Monsieur G died "after a painful illness which he endured with courage." The death certificate lists his next of kin, including a "Monsieur Daniel Pauli [sic]." With the death of the father, any semblance of a family home disintegrated rapidly—the G--- sons divided up their old man's business and became regular customers at the local criminal court. Meanwhile, their mother set up shop at a neighborhood bar, where she acquired a taste for televised wrestling matches.

Daniel found himself out in the cold once again. This time, it was Suzy who offered him refuge. By now, they had known each other for over a year. He was more authentic than the rest of their classmates, whom she found boring, especially the girls. Like two peas in a pod, they would wander through the city, talking about all kinds of things, taking their time walking each other home. Daniel carefully avoided the G---s and their friends, preferring to follow Suzy back to her parents' bakery. The parents in question were not happy about their little romance, but the kids used the infallible excuse of shared homework. The nice thing about a bakery is that everything is within arm's reach. Daniel could get a bite to eat and tease Suzy about her blond curls and her ballet and piano lessons: "You're so bourgeois!"

But even with good bread and Suzy's pretty eyes, Daniel's situation was still unstable. He was penniless and almost living on the street, and he couldn't count on Suzy's family to take him in. At school, things were deteriorating rapidly—his absences piled up and his grades from December 1962 were fatal: an average just above the minimum, but with two failing grades, in math and English, as well as particularly low scores in Spanish and stenography. Daniel was expelled with the following comments: "Inadmissible conduct including numerous instances of tardiness and negligence." At the beginning of 1963, he worked in the watch industry for a few months, at Incabloc. But he'd had enough of this town in the middle of nowhere and its six months of winter—it was time to move on. Daniel said goodbye to Suzy. She knitted him a new sweater, which he wore throughout his wanderings in the years that followed. He left her a portrait of a young man without a smile.

"It breaks my heart," Suzy says. "He succeeded, of course, but for me, there's always a sadness in his eyes."

YOUTH IN GERMANY

I N JULY OF 1963, a young man of seventeen, tall and skinny as a beanpole, turned his back on La Chaux-de-Fonds and Switzerland. Baptized Catholic, Daniel had only a few months earlier sought out a priest who had finally given him his first Communion, and in Germany an Evangelical charity* in Mönchengladbach, near Düsseldorf, took him in. The Bundesliga had not yet come into existence and, like most German cities, Mönchengladbach was still rising from the ashes. Two-thirds of the town had been destroyed by 65,000 incendiary bombs. In that same year of 1963, Konrad Adenauer, the first chancellor of the Federal Republic of Germany, was finishing a fourteen-year term largely dedicated to reconstruction. This economic miracle bore fruit, and West Germans entered into a frenzy of consumerism and international travel that has continued to this day. Very few Black people lived in Germany at the time; they had been persecuted and killed by the Nazis in the same way as the Jews and the Romani, and suffered from a public image straight out of a colonial exhibition. "Who's Afraid of the Black Man?" was a popular children's game at the time, and, until the 1990s, the term *Neger* was still part of every-day language. During my studies in Kiel, I struggled to explain to my German colleagues that the word was really too close to its horrible French and English cognates. Yet, even if he was occasionally the target of verbal racism, Daniel never experienced any physical violence and found Germany "friendly and welcoming" overall. The Black people

19

---

* Diakonisches Werk.

who settled in East Germany or who live in Germany today have not always been so fortunate.

Daniel's plan was to go to Germany to perfect his language skills, then return to La Chaux-de-Fonds to work as a translator. Even if he was running away, it was not a wild escape; Daniel traveled with his French passport in due form. Like Candide, he arrived in North Rhine-Westphalia, where the Evangelical mission offered young people the opportunity to do a year of charitable service. In Mönchengladbach, he began working at an institution for mentally disabled people run by the Lutheran Church and helped care for a group of young male patients. He was fed and housed, given blue scrubs and seventy deutsche marks' worth of pocket money each month. His German turned out to be too academic, the result of book learning, but he mastered the language quickly and would later read and write it perfectly. European psychiatric hospitals had not yet undergone the revolutionary transformation of the 1970s, and his new work environment was oppressive and entirely masculine. His coworkers praised his good humor, however, and he often played the clown to help cheer up his patients—a group of twenty or so adolescents and young adults, some of them with Down syndrome, others with hydrocephaly, or victims of accidents like one young man, saved too late from drowning, whose angelic features always remained perfectly expressionless. Patients who made trouble were given heavy-handed treatments that left them dazed and confused. The neighboring group was run by a portly man whom many suspected of abusing certain patients.

Young Daniel's burgeoning faith did not survive his six months in the Lutheran Church's institution. Daniel recalls one patient in particular who suffered from a serious developmental disability: "His body and face had no consciousness behind them; he had grown to resemble a wooden plank, the outline of a being without any depth, his functionality limited to his vital organs." Someone explained to Daniel that God had deliberately made the boy this way in order to inspire empathy in others. Daniel was outraged by the claim that a supposedly benevolent God would use a human being as a tool in that fashion. He found that kind of manipulation totally unacceptable, a sign of divine contempt for the human condition. The church of the time also took an odd approach to managing the sex lives of its young seminarians: those who were in training to be social workers were not allowed to sleep with members of their flock for the first two years. But during year three, they received what everyone called their "hunting license."

DANIEL LEFT THE asylum, and the Kingdom of God, in early 1964. After a monthlong training course in first aid, he finished the second half of his charitable service as an orderly in a hospital in Wuppertal. He was happy to be in mixed company again, and one Sunday, he overheard a group of girls talking about their approaching baccalaureate exams.* Their words hit Daniel like a freight train—he suddenly felt trapped, without a future. Over the next few days, he asked around and learned about some night classes where young people who were already working could prepare to take the baccalaureate exam, which would then allow them to go on to higher education.

The initiative was associated with a press campaign that encouraged working-class families to keep their children in school.† Daniel spoke with the director of the program, who agreed to admit him despite his foreign origins and still imperfect German. "You'll definitely learn something," the director told him at the end of the interview. Daniel waited until classes began in autumn of 1964 to leave the hospital in Wuppertal and go back to school. But he still had to put bread on the table, and he started looking for a new job with hours that wouldn't overlap with his classes, which were from 5:00 PM to 9:00 PM, five days a week. At a time when there was full employment, and in the center of one of the largest industrial districts in Western Europe, nothing could be easier.

---

* The baccalaureate is an academic qualification that students are required to pass in order to graduate from high school in Germany, France, and other European countries. (TN)

† With the slogan *"Schick Dein Kind länger auf bessere Schulen"* ("Send your child to better schools for longer").

Daniel found work at Herberts Lacke, a paint factory where his shift ran from seven in the morning until four in the afternoon. The main product was automobile paint, and his employer ran a research laboratory next to the production line. Daniel worked there doing quality control on raw materials using large quantities of various solvents. Because he was not a real apprentice, he was given only the most repetitive tasks under the unfriendly eye of the lab manager, an alcoholic *Herr Doktor* who sometimes burst screaming and shouting out of his office. They worked without any kind of protective equipment, and the chemicals left over from testing were poured down an open sewer that dominated the center of the room. Daniel's nose bled regularly and abundantly. The big boss, Kurt Herberts, was, however, a judicious man who had distinguished himself during the Nazi period by commissioning paintings from "banned" artists to help them survive. A follower of the theories of Rudolf Steiner, he founded two schools and developed a progressive social policy within his business, establishing a training center for three hundred apprentices. Daniel benefited, albeit indirectly, from Herberts's enlightened capitalism: the lab manager who replaced the noisy alcoholic left Daniel alone and even gave him permission to do his homework during business hours.

Daniel's night classes followed the German high school curriculum of the period, with all the classic subjects, including Latin. One hundred and eleven students were divided into two classes; four years later, only twenty-five of them sat for the baccalaureate exam. Daniel slept very little between the factory, his classes, and his ever-accelerating bibliophagy. At eighteen, one cannot really do without friends, and Daniel saw his when he could, usually late at night. He rented rooms from a series of "old harpies"—one of his landladies was even afraid he would stain the sheets with his dark skin by sleeping naked. Tough negotiations were required if he wanted to have any visitors, male or female. The winters in Wuppertal were a little less frigid than the ones in La Chaux-de-Fonds, but most German homes at the time lacked central heating. The young worker-student's room was equipped with the most sluggish and dusty of coal stoves, and heating his attic apartment proved impossible when he regularly came home after ten o'clock. Daniel left his door open in hopes of catching some heat from the rest of the house while he did his nightly reading—but it was quickly closed again by his landlady. He lived in a state of constant fatigue but pushed through it. With no family to wake this almost adult each morning, he worked without a safety net. Daniel had read Sartre, though, and he took an existentialist approach: once he had decided to get his baccalaureate, nothing could stop him.

Daniel's most important encounter at the end of 1964 was with Walter Kühhirt, who also worked at Herberts Lacke and lived in the same boarding house on Emilienstraße, not far from the Wuppertal-Barmen train line. Walter was a white African, Daniel a Black European. Both foreigners in Germany, they hit it off right away. Walter was born during the bombardments, near Düsseldorf. His father, a German who grew up in colonial Namibia, fought in the Afrika Korps during the war. He made it out with his life and took his family to live in Windhoek* soon after. Walter was six years old at the time and discovered the old "South West Africa."† His African childhood would influence him for the rest of his life. Walter's family was mostly made up of missionaries, compassionate but straitlaced people in a country suffering under the yoke of implacable racial segregation. His white enclave on the edge of Windhoek bordered on a Black ghetto that bore a colorful name meaning "old shipyard."‡ Like District Six in Cape Town, South Africa, this neighborhood was soon emptied by force in order to make room for the expanding white areas. The Black inhabitants were gradually rehoused in the infamous neighborhood of Katutura ("the place we do not want to live" in the Herero language). Walter clearly remembers that night in December 1959 when the "old shipyard" was invaded by soldiers who beat and burned,

---

* Now the capital of Namibia.
† South West Africa was a German colony between 1884 and 1915. After World War I, it became a mandate of the Union of South Africa, and in 1990, gained its independence as the Republic of Namibia.
‡ *Die alte Werft.*

killing thirteen people. The acts of Black resistance in response were, however, a seminal moment for the SWAPO.* Walter would never forget the blazing sky and the sound of gunfire, nor the moment he saw his father get down his old Mauser to guard the door.

In 1963, Walter was nineteen years old and handsome like a young Steve McQueen. His father had sent him back to Germany to "learn to live independently," an authoritarian decision that felt like rejection. A grandmother was supposed to take care of him, but he wound up in a boarding house for young workers with four to a room. In the winter of 1963, the Rhine froze over, and Walter suffered from the cold—he never would get used to it. City life was also a shock to him. When he rode the tram for the first time, the motion of the vehicle knocked him over. He explained to the other passengers that he had just come from Africa and had never seen a contraption like this before; they gave him odd looks. Forever an outsider in his native country, he would always have a soft spot for displaced persons.

When I meet him in the train station in Düsseldorf one night in January 2016, this kind, soft-spoken seventy-two-year-old man speaks to me at length about Daniel, but also about the one million refugees that Germany has recently taken in and whose future worries him greatly. Walter gives literacy courses to migrants and he reminds me of my own father, who does the same in France. I feel bad for asking him to meet me in a freezing train station so late at night. We finish our interview in my hotel room, where he also asks me about my work, then I walk him back to the station in the rain. On the way, he tells me that he is not afraid of dying, just of losing his mind. I sense that he appreciated tonight's recall exercise but found it somewhat disturbing as well. This would be the case with many of Daniel's friends and family members during my research.

But in the 1960s, Walter was the life of the party and Daniel was not to be outdone. They didn't talk much about their painful childhoods, preferring instead to discuss the different girls they'd met.

---

* South West Africa People's Organization, the party currently in power in Namibia.

Walter is the only man with whom Daniel would discuss his emotions freely, a degree of closeness and mutual trust that the pair rediscovered forty-five years later during a short vacation on the North Sea coast. In the Wuppertal of their youth, daily life was marked by a turbulent yet joyful pauperism. Always on the lookout for free entertainment, they joined a union for foreign students. Walter had read Bertrand Russell, and the two friends soon developed a reputation as atheist trouble-makers. They had the most fun in a discussion group about religion. During one meeting, each person was supposed to talk about why he believed what he did: an easygoing Ghanaian explained that he was Muslim because that's how it was in his family and quickly caught flak from his Iranian classmates for his lack of fervor.

Soon, the winter ice gave way to a gray drizzle, which then evaporated in the summer sun—it was time for a vacation, and Daniel and Walter decided to hitchhike around Europe, like a whole generation of Baby Boomers. During the summer of '65, they set their sights on England, then the next year, on Sweden, which Daniel absolutely insisted they cross from end to end in order to visit a girlfriend of his in Finland. The girl in question would wait a long time—the Nordic roads were so deserted (or the drivers so racist?) that they never made it out of the southern half of the country. After three days of waiting on the side of the road, the duo threw in the towel and beat a slow retreat back to Northern Germany. Just back over the border, they were getting ready to sleep under the stars yet again when they encountered a group of police officers who took pity on them: the jail cells down at the station were empty, and they offered to let the travelers stay the night. Daniel's first encounter with the law was quite funny... and very unlike the one that awaited him back in France.

IN FACT, DANIEL'S relatively calm life in Wuppertal was soon to come to an end. Even if he rarely talked about them, Daniel had not for-gotten his French origins. Looking for his mother, he wrote to the public authorities, but never received a response. Then, in 1965, the French army tracked him down through the embassy in Bern with some help from the German government, which requires all residents to declare their address: at nineteen years old, Daniel, like all young Frenchmen his age, was called up for a year of mandatory military service. He ignored the French government's first letter—they had never done anything for him, so why should he respond? The pastor at his high school in Wuppertal, putting Daniel's increasingly obvious atheism aside, offered to help and tried to obtain a dispensation that would allow Daniel to finish his studies. Nothing came of it, though, and at the end of 1966, things went south—he was declared a draft dodger and lost his French passport. Left with only a laissez-passer, he was ordered to return to France immediately. Around the same time, the army barged into his mother's home in Paris, searching the apartment twice under the fearful eyes of brothers and sisters who were still unaware of their eldest sibling's existence. Renée Pauly, who had still not been told that her son was in Germany, wrote to Madame G's sister, begging her to find Daniel. The woman had moved from Paris to La Chaux-de-Fonds, and though she had no idea where Daniel was, she kept the letter.

Around Christmas of 1966, Daniel put his studies on hold one more time, left his job at Herberts Lacke, and returned to France via La Chaux-de-Fonds, where he still had a few friends. He ran into the G--- aunt and she gave him his mother's letter, which turned his world upside down. He arrived in Paris on a Friday in January of 1967, only twenty years old and already leaving several lives behind him.

Daniel does not go into detail when discussing his reunion with his family but admits that everyone cried a lot. His brother Gérard, the third oldest of the siblings, who was seventeen at the time, is more forthcoming. When I visit them near Aix-en-Provence in February of 2016, Gérard and his wife, Jocelyne, have just come down from their home in La Rochelle to look after their grandchildren. Up to this point, I have done most of my interviews in noisy public places, and I appreciate the calm of their rented cabin in the countryside. Gérard is a retired house painter and maintenance man, tall like all the Paulys, and athletic, with a kind voice and eyes.

Daniel's arrival in Paris, as Gérard describes it, was a family fiasco. Renée revealed the existence of Daniel to her other children at the last minute. On the day of, she left with two of her daughters to wait for him at the Gare de l'Est, leaving her husband at home with the rest of the clan. They missed each other at the train station, and Daniel found his own way to the family home in Maisons-Alfort, where he introduced himself and asked if this was the Pauly residence. They invited him in and he sat down to wait, grim-faced and stiff as a board. Daniel had a list of very specific questions designed to determine if, yes or no, he had really found his French family. The ambiance remained tense until Renée arrived and locked herself in the kitchen with her adult son. She showed him the letters from the G---s, proof of their maneuvering, lies, and threats, as well as of her own efforts to find him. We can only imagine how strange they must have felt, reunited at last, joyful and angry all at the same time.

But that was only the beginning: the following Monday, Daniel made his way to the army barracks, where he was arrested and put under guard like a dangerous criminal by two soldiers armed with submachine guns. Torn from his mother and thrown in prison, Daniel lived through a terrifying week. He was completely lost in that big, man-grinding machine; he "didn't know the first three notes of 'La Marseillaise,'" saluted with his left hand, and, when entering a colonel's office, politely asked if he might have a seat—after which everyone yelled at him. But once again, someone reached out a helping hand: Daniel was saved by an army psychologist who recognized that they weren't going to get much out of this born nonconformist. He offered to declare Daniel mentally unfit for service, which would leave a permanent mark in his file but save his skin. Daniel accepted and was out by the end of the week. Four months later, he went to trial for dodging the draft. A public defense attorney he'd never seen before represented him half-heartedly. The military judge was, Daniel remembers, "a big Black man from the French Caribbean dressed in a red robe," who let him off with a two-month suspended sentence—after all, the reasoning went, he was the eldest of eight and belonged at home. The Jehovah's Witness and conscientious objector whose case was heard just before Daniel's did not have the same luck: he had already been behind bars for two years and was given three more, the whole affair wrapped up in about five minutes. Relieved, Daniel told himself that the Reign of Terror* wasn't just a thing of the past.

Freed from his military obligations, he still had to wait for a new passport before returning to Germany, and the whole adventure cost him another school year. In the meantime, he stayed with the Paulys, who squeezed a little tighter into their apartment, where there were now ten of them, to make room for Daniel. He found work at the DDE (Department of Transportation and Infrastructure), evaluating the traffic exiting Paris on the east side to help prepare for the construction of the A4, France's second-longest autoroute. He enjoyed counting cars in the field, but not the tedious task of inputting data on the enormous mechanical calculator.

During evenings at home, Daniel's mother, "a real chatterbox," told him all about the family history of the Paulys and the Cléments. His brothers and sisters got to know him and were soon fascinated by this intelligent and cultured young man who shared their mother's talkative nature. Daniel's visit left a strong impression—he was a window on the world for all his siblings. Work began and ended early at the DDE, so Daniel took advantage of his afternoons off to explore Paris with his brothers and sisters. Barefoot, his nose in a book (preferably a tough read†), he would ride the metro with his siblings, then guide them through the Louvre and the Palais de la découverte. On weekends, they built a whole collection of photographic equipment, including a waterproof housing for a camera that could take pictures underwater, buying raw materials at the flea market and putting them together under Daniel's watchful eye. He had learned

---

* Reign of Terror (1793–94): One of the darkest periods of the French Revolution, during which frivolous accusations of treason led to multiple massacres and executions. (TN)

† *Ulysses* by James Joyce was one of the books he read around that time, and he would gladly go on and on about the famous chapter that lacks punctuation.

a thing or two about mechanics and assembly at Incabloc and from his years of building model airplanes. His two youngest brothers, Gilbert and Christian, developed a passion for photography that would stay with them all their lives. For his part, Daniel was favorably impressed by Louis Pauly, a "good guy" who had married Daniel's mother and recognized him as his own. The two of them got along right away, and Daniel admired this new father, who woke up early each morning to fight the good fight as a heavy industry worker at the metallurgy company Tréfimétaux. Louis's heart and his ballot leaned left, and he harbored a distrust of intellectuals that Daniel would never forget.

For summer vacation, the ever-loyal Walter and another friend came down from Germany to perfect the tableau. Walter was a wild one and shocked the Paulys right off the bat by imitating a whole host of different characters. Redbeard the pirate and the British comedian Marty Feldman were among their favorites, though it's hard to imagine how Walter managed to imitate Marty's bulging eyes. Amid the collective frenzy, Daniel continued digging into his family history. He bought a scooter and went alone to meet Renée's father, Henri Clément, the man whose racism had kicked off this whole psychodrama. In Avize, his grandfather received him coldly but received him all the same, and Daniel returned to Paris with another item crossed off his emotional checklist. Eventually, the money he made counting cars allowed him to trade his scooter for a small three-horsepower van. Daniel's German girlfriend, Ute, had not forgotten about him during his stay in France, and came to join him for a trip to England. The Paulys thought she was a little quirky, but very sweet.

At the beginning of the 1967 school year, Daniel returned to his night classes in Germany, but Herberts Lacke wouldn't have him back— after all, he was only going to leave again after graduation. He was forced to take a much more tiresome job in a brush and broom factory. The owner was appalling and Daniel found new employment as soon as possible, this time as an office-furniture deliveryman. The social exclusion he experienced, his life as an adolescent worker, and his reunion with his working-class family naturally pushed Daniel toward Marxism, and it was during these years that his political views began to take shape.

IN 1968, STUDENT protests were heating up all over Europe. In May, Daniel and Ute loaded up an old Volkswagen with extra fuel (most gas stations were closed because of labor strikes) and headed to Paris to see the revolution in action. With Daniel's Pauly siblings Gérard and Anita, they participated in four or five demonstrations, mostly as spectators, amid occasional onslaughts from the riot police and the metallic smell of tear gas hanging in the air. Troublemakers mixed with the protesters, "throwing cobblestones a lot further than the students could have." These games of hide-and-seek contrasted sharply with evenings at the Pauly residence, where the family often wondered how they would manage to feed everyone the next day. Daniel also took Gérard to the Odéon theater, which was being occupied by student protesters. Getting in was no easy feat, but Daniel passed himself off as a German journalist accompanied by his "assistant," who didn't say a word. Gérard was struck by the beauty of the place and by the content of the discussions going on there, in which the revolutionaries were already assigning themselves cabinet positions in their future government. Daniel, however, was underwhelmed by their Maoist diatribes, which he found childish and... well, theatrical. He couldn't resist standing up to tell them about how his German comrades were organizing. Some of them had, in fact, declared war on "the establishment" the year before. In November 1967, during a public university meeting in Hamburg in the presence of the city's crème de la crème, two students took advantage of the university rectors' arrival to roll out a banner that said, "Under their gowns, a thousand-year-old stench."* The reference to the thousand-year Reich and the Nazi past of some postwar faculty was clear and would be remembered by posterity—the student revolution in Germany had found its battle cry.

---

* *"Unter den Talaren, der Muff von 1000 Jahren."*

In his last year of high school, Daniel received a scholarship from the German government, and his life took on a less frenetic pace. To avoid losing his edge, and because stagnation made him nervous, he taught private French classes part-time for a few months, though he didn't particularly enjoy it. When spring brought the baccalaureate exams, Daniel did well overall but panicked on the math test and received a mediocre score.* He was mortified, and a long way from imagining that later he would publish volumes stuffed full of equations. The local newspaper in Wuppertal covered the event with the headline, "No special treatment for late graduates."† The article, from the spring of 1969, specifies that twenty-four of the twenty-five candidates, with an average age of twenty-five, passed the nine written tests and almost as many orals. Only a few of the candidates were interested in going on to higher education. A single student is named in the article, one Daniel Pauly from Paris, France, who declared that he had paid for his own schooling and wanted to go on to study biology. Daniel was also photographed for the article: he appears thin, with short hair, a white button-down shirt, and a skinny black tie. He is deep in conversation with his classmate, Sabine Jensen, a pretty young woman in a black vest who is running a nervous hand through her white-blond hair. Daniel's hands are stretched out in front of him, palm to palm, fingers spread.‡

His high school diploma in the bag, Daniel was thrilled to finally move on to university, but before that, he still had a few skeletons to chase out of the closet. In June of 1969, he set off for the United States in the company of the faithful Walter to search for his father, Winston McLemore, and his African American heritage.

---

* Four out of six, with one being the best.
† *"Senior"–Abiturienten wurde nichts geschenkt–Mit Cicero, Algebra und Goethe zur nachgeholten Studienreife.*
‡ A gesture also often made by Barack Obama, another mixed-race boy who grew up without his father, in Hawaii.

THE SEARCH FOR
AN AMERICAN FATHER

"I GREW UP IN the shadow of Emmett Till, who was lynched in Mississippi at the age of fourteen. That young man was slightly older than I was." Ed Whitfield—sixty-seven years old, six feet two, sturdy, with a white beard and shaved head—speaks to me about his life and that of his family in racist America.

Ed is Daniel's cousin, the son of his American aunt, Winifred Whitfield. We are seated comfortably in the offices of the Fund for Democratic Communities, a foundation that he co-directs in Greensboro, North Carolina. Before the recession, this building held the largest Wrangler jeans factory in the nation—the floor still bears traces of the workers' shoes and the equipment that once stood here. In the distance, a freight train blows its whistle as it crosses a countryside draped in warm autumn colors.

I came down from New York in a tiny plane to meet Ed at the Greensboro airport, which, on Thanksgiving Day, was totally deserted. Good Christians were home with their families and Donald Trump had just been elected president of the United States. After a few hours of suspense, Ed appeared—he had just returned from a visit to his daughter in Atlanta. I am happy to have caught up with this rare bird, but for the first time in my investigations, I feel almost like an imposter, an insignificant white man lost in the Deep South. If I believe what I hear on the news, nothing ever changes in America: the police still shoot Black people in the back with impunity, the prisons are full of brown people, and the rift between communities doesn't seem to want to heal. Meanwhile, Trump is well on his way to joyfully pulling apart everything Obama worked so hard to build over the previous eight years. "Trump's election is going to encourage the racists," says Cousin Ed. "In the South, the KKK mostly celebrated his victory."

Daniel's American family is from Little Rock, Arkansas, a place where 284 Black people were lynched between 1883 and 1959, with the tacit agreement of the police and the public authorities. To hear Ed tell it, the Black community reacted to this ambient racism in two ways. First, by arming themselves. "When John Carter was lynched in 1927, the mob started fanning out into other parts of the city looking for some more Black folk to mess with. I understand from some historical reports that they only stopped when a few Black people came out on their porches with shotguns to say, you know, don't bring that mess up in here. My father was eighteen years old at the time. I grew up in a house full of guns, and it just seemed normal to me. My sister had a .410 shotgun and my brothers and I had .22s. My father had guns and pistols. And he wasn't a hunter or anything, but he would take us to target practice and talk to us about gun safety. The first time I heard anything about race was when my parents told me that we would go downtown and I needed to be careful because there would be white* people. I was quite young, and I thought, 'White people, that's interesting!' Our walls were kind of a green color, and the door frames were white, and I was thinking to myself, 'I've never seen people that color.' I was three or four. When I saw them, I was really disappointed—'I thought you said *white* people!'"

The second way the Whitfield family fought for their rights was through academic excellence and the desegregation of the education system. Ed's mother, Winifred, was an exemplary figure in the fight for civil rights who died a centenarian in 2015. She became a teacher in 1929 at the age of seventeen and eventually earned her master's degree in education. All of her children went to "the white school."

---

* Ed has a way of pronouncing "white" so that you can hear the "h."

"People think that the reason for school desegregation was because you had all these poor Negroes who had these terrible schools in their community. That's not it. In Little Rock, the education that was available to us in the predominantly Black schools was a quality education," Ed says. "Opportunities were much more restricted, so there was nothing else that a smart Black person could do but become a schoolteacher. It was really about proving that we were as good as they were. For a lot of those young people, it was pure hell. My sister went to Little Rock Central High School in 1960—this was the school where the president of the United States had to send troops to in 1957 to allow those nine Black children to go to school. But she went, and I remember hearing her stories about being elbowed in the hallway, having ink poured on her in the classroom, having soup spilled on her in the cafeteria. She wrote a poem about it. I still remember the first lines:

> I walk these great halls of Central High School
> The treatment I get is none for a fool.

Ed went to the same high school in 1964. He was tolerated, barely, though he wasn't allowed to act in the senior class play "because there were no roles for Negroes," nor did he participate in an exchange program with a school in Madison, Wisconsin. "We couldn't find a black child there to exchange with you," they told him. Ed never made a grade below an "A" except in music, where the grading was subjective and the teacher racist, though that didn't stop Ed from playing Handel's *Messiah* on his oboe for the Christmas concert at the local Black college. But Ed's particular gift was for mathematics. In 1965, he spent the summer in Ohio taking advanced classes, where he excelled in group theory. He spent the next two summers in a program at Cornell University, first as a student, then as a tutor. His rise to the top was astonishing: in the summer of 1967, at age eighteen, he was invited to the White House to receive the US Presidential Scholars medal from Lyndon B. Johnson. In the official photo, young Ed is slightly taller than LBJ. "I wanted to refuse it. At the time, I was already active in the NAACP* youth chapter and very much opposed to the war in Vietnam. And then I thought about it and I realized, if I do this [refuse the medal], my mother will kill me. That's one thing that I kind of regret. I would have been the only person ever to do that."

Ed would soon climb even higher. The presidential scholarship opened doors for him at all the major universities. He received a very serious offer from Harvard but eventually settled on Cornell, where he already felt at home. He took classes in mathematics, philosophy, political science, and African American literature in a program designed to propel promising students from a bachelor's degree to a doctorate in six years, top speed.

But history would decide otherwise. During his sophomore year, Ed became president of Cornell's African American student union and quickly suggested changing their name to the Black Liberation Front as an homage to the Viet Cong. This new political engagement came at a time of extreme tension: Martin Luther King had just been assassinated, and the Tet Offensive had marked a turning point in the Vietnam War. Early in 1969, African American students at Duke University in North Carolina occupied the administrative buildings to attract public attention to the issues they were facing. The Duke protests inspired Ed and his classmates at Cornell, who had a few complaints of their own: Black students who straightened their hair in the dorms, creating a smell unfamiliar to their white classmates, were being accused of smoking joints. And worse, some "jokers" with ties to the KKK had burned a giant cross on campus, a traumatic sight for young people from the South. Lastly, after participating in a protest, a group of African American students had been brought before a disciplinary commission organized by the university that was entirely made up of white men.

Ed and his friends decided to stage an occupation—an armed one, whose objective was to create as much disruption and attract as much

* National Association for the Advancement of Colored People.

attention as possible. They achieved their goal: a reactionary historian later called it "the day Cornell died."* On April 20, 1969, Ed's heavily armed team took over Cornell's Willard Straight Hall. "It could have gone very differently, but the university president was a Quaker, a guy named James Perkins, and he was dead set on not wanting anyone to get hurt," Ed recalls. Perkins would be harshly criticized for what his detractors called a permissive attitude, but his actions likely prevented a massacre. He agreed to the revolutionaries' demands and offered to take their weapons discreetly out the back door before they gave themselves up. Ed refused: "We came in here with 'em, so we're going to go out with 'em." The protest was a windfall for reporters, especially one Steve Starr, who won a Pulitzer Prize in 1970 for his photo of Ed Whitfield leading a group of protesters out of Willard Straight Hall. Ed hardly looks like the same young man who had gone to the White House in his Sunday best twenty months earlier—he has long hair and a beard, and he carries a rifle in his right hand. "I am so happy that we didn't have to do anything with those guns other than have them," he says. Those images appeared all over the world: Daniel saw them in Germany, and his French family discovered them in *Paris Match*.

Ed's career at the university was over. He wasn't expelled and would even return to Cornell the following semester, but his heart was no longer in it. He taught for two years at Malcolm X Liberation University in Greensboro before going to work in manufacturing, where he was a union representative for over thirty years. Since his retirement in 2009, he has been a spokesperson for the Fund for Democratic Communities, which works to create "a system which is not exploitative, based on reconstituting the commons."

I ask him if he regrets leaving behind the university career that could have been his if he had been willing to toe the line, like most of his classmates. "That would be so boring to me," he says. "I would have a hard time living with myself under those conditions. My ambition in life was never to try and figure out if I could make myself successful.

---

* Thomas Sowell, "The Day Cornell Died," *Hoover Digest*, no. 4 (1999).

I was always trying to figure out how I could bring my community along, and that's why, fifty years later, I'm still marching, even if I'm not as quick as I used to be. But," he adds with a smile, "I'm working on my second retirement as a famous blues musician." A beautiful bass guitar is in fact sitting on a pile of things in the back of his car.

During the 1950s and '60s, Winifred Whitfield managed to stay in touch with Renée Pauly. "My mother talked about Daniel a lot," Ed recalls. "She just felt bad about what her brother had done, going over there and having a child and not doing anything to take care of him. She thought that was awful." Winifred sent packages to Renée for Daniel, and it was through her that he eventually contacted his American family.

UPON ARRIVING IN New York at the end of May 1969, Daniel was immediately pulled out of line by customs agents, the first stop and search of many. These gentlemen, who were oddly convinced that Daniel was a drug smuggler, seemed fascinated by his orange-and-blue-striped socks. Walter waited for him at the exit, and Uncle Carl (Winston and Winifred's brother) and Cousin Robert (Ed's older brother) picked them up from the airport. Giants, like all the men in their family, they loaded Daniel and Walter's luggage into an equally giant American car. The two friends spoke a little English, but they had a hard time understanding Daniel's uncle and cousin, who drove them toward New Rochelle.

Uncle Carl owned a garage, a taxi service, and a number of other more-or-less-legal businesses in Harlem. Thin and elegantly dressed in black, the forty-year-old projected a relaxed charm that Daniel liked immediately. Carl was also a collector for an illegal numbers game, and he invited his European nephew to come with him on his rounds. They ended the day in one of the neighborhood's basements. Outside, an Irish policeman twirled his nightstick while pacing the sidewalk, just like in the movies. "The cops were incredibly corrupt—that one was looking out for his bread and butter," Daniel remembers. "Another big guy was guarding the entrance; Carl told him that I was 'OK.' The inside was dimly lit, everyone was Black, everyone was smoking, the guys who ran the business wore dark sunglasses. Working men went there to lose their money playing cards or dice or other games I wasn't familiar with. The tables were covered with wads of cash, and whole piles of it flew from one end of the room to the other at high speed. It was an amazing thing to see, kind of an organized chaos, very picaresque."

Daniel and Walter worked at Uncle Carl's garage for a while, then took a Greyhound bus to Little Rock, Arkansas, where Aunt Winifred and the family were waiting. "He looked like my mother's family," Ed says, remembering Daniel's arrival. The young Frenchman was in awe of Ed and his friends—"Black American revolutionaries armed to the teeth"—and by "the enormity of what had happened at Cornell." An unmarked FBI car followed Ed everywhere he went, and Winifred strongly recommended that Daniel not run in the street if policemen were present. Ed and Winifred told him all about the family history, the fight for civil rights, and Black consciousness. Ed spoke about his "undying love for African people." "I didn't understand at the time how meaningful that visit was for Daniel," Ed recalls. "Later on, he wrote a letter thanking me for spending time with him and talking about how he had really found something he could be a part of." Meeting his family and seeing a Black America that was fighting for its civil rights at the end of the 1960s was like an electric shock for Daniel's politics and identity.

Over the next three months, he and Walter traveled around the US on Greyhound buses, thanks to the 99/99 ticket, which, for ninety-nine dollars, gave them unlimited access to the whole network for ninety-nine days. They crossed the continent (three days and three nights, nonstop) because Daniel's father, Winston McLemore, now retired from the US Army, lived in Seaside, California. Winston and his wife, Mary, welcomed Daniel into their home, but things didn't click the way they had with the Whitfields. Winston voted Republican and his worldview was diametrically opposed to that of his son. After a few days of living with his biological father, Daniel discovered that

Winston was a serious alcoholic whose episodes sometimes led to violent altercations with his wife—everything Daniel had learned to hate in La Chaux-de-Fonds. Mary was also suspicious of Daniel, convinced that he had come to claim his share of her husband's inheritance. Winston, an honest enough man despite his standoffishness, offered to help support Daniel while he finished school. "Fifty dollars* a month would really help," Daniel suggested. "I could go to school without having to work nights." Winston agreed, but Mary made sure that Daniel never saw a penny. "Meeting his father must have been a big letdown," Walter says. "But he didn't show it. We clowned around like always, even put on a little show for Winston and Mary."†

The duo didn't stay in Seaside for long, choosing instead to go mix with the flower children at Berkeley. Then, at the "Celebration at Big Sur," they discovered the music of Joan Baez and Joni Mitchell amid clouds of marijuana smoke.

After two decades of waiting, the visit to Seaside was a bitter disappointment for Daniel, but it did lead to a much more significant encounter. The McLemores were friends with a half dozen other Black army retirees, including the Wade family. Mrs. Wade—who had grown up with Winifred Whitfield and, like her, knew how to run things—sent her grown daughter to chauffeur Daniel around during his stay. And that's how Daniel Pauly met Sandra Wade—but we'll come back to her later.

In the meantime, Walter and Daniel continued their road movie on the Greyhound, passing the time by playing endless games of chess, which Walter almost always won. In August, they ended up in New Orleans, where they witnessed the evacuation of the city ahead of Hurricane Camille (from which it was ultimately spared). Penniless as usual, they took a tour through New Orleans's underground. They finally realized that, to the people they met there, a Black-and-white

---

* Fifty US dollars in 1969 is about 320 dollars in today's money.
† "There's no point, there's no sense in having a father you literally can't communicate with, whose feelings for you have always been open to question." Marie Ndiaye, *Three Strong Women: A Novel*, trans. John Fletcher (London: MacLehose Press: 2013), 12.

duo could only be a gay couple. This prejudice would be confirmed by Ted Conover during his study of American hobo culture a decade later.* Daniel described their picaresque adventure in a humorous text entitled *An Innocent's Summer.*†

He and Walter were a lot less innocent, however, when they boarded the plane back to Europe in New York. Walter, fascinated by the States, applied to one American college after another, eventually completing his doctorate in applied chemistry in Fargo, North Dakota. Daniel, for his part, could no longer see himself having a relationship with a white woman and broke up with Ute as soon as he returned to Germany. "I was terribly upset by everything I had seen and I reacted badly," Daniel admits. "I know now that what we had was worth more than my convictions. She was a wonderful girl, and she helped make me into an open-minded adult."

---

* Ted Conover, *Rolling Nowhere: Riding the Rails With America's Hoboes* (New York: Vintage, 2001, first published 1984).
† Unpublished.

II

CONSTRUCTIONS

FROM OCEANOGRAPHY TO FISHERIES BIOLOGY

S EPTEMBER 1969. DANIEL found himself in Kiel, Germany. He had decided on a coastal town, with the romantic idea that life would be sweeter next to the sea, and a major in agronomy, since the world's people needed to be fed.

Kiel is the capital of Schleswig-Holstein, a state situated so far north within Germany that for much of its history it was part of Denmark, and it still bears the scars of the last Ice Age: it is a moraine-covered region along the Baltic Sea, itself a vast glacial lake that freezes readily in the winter. Since Otto von Bismarck's time, Kiel has also been the headquarters for Germany's navy. It was a beautiful Hanseatic city destroyed by Allied bombs, then destroyed again by the uniform, "modern" architectural style of postwar reconstruction.

In Germany as elsewhere, the upper echelons of the navy were a reactionary bunch. The few people who hoisted a white flag as the Allies approached Kiel were shot by their neighbors. After the war, a good number of Nazis lived out their days in the countryside of Schleswig-Holstein, often occupying high-level administrative or political positions. Daniel knew about all this from his reading, but he would soon experience it more directly: on a local fish farm where he worked one summer, his boss told him casually that he was taking the afternoon off to go celebrate the *Führer*'s birthday with some friends. A quarter century after the war, the agronomy department at the University of Kiel was still packed to the gills with old Nazis. The young students suffered from harassment, and the atmosphere was awful. Daniel learned that this port city was also home to an internationally renowned oceanographic institute, which was recruiting an army of young professors with more liberal ideas and offering courses in ocean science. As unlikely as it seems, after growing up in landlocked Switzerland without any special interest in marine environments, Daniel became a student oceanographer. In a way, those old Nazi agronomy professors gave marine ecology one of its most emblematic figures of the twentieth century. Things could have easily turned out differently: Daniel the agronomist, Daniel the doctor, Daniel the jurist, the linguist, the orchestra conductor—the possibilities are dizzying.

Germany isn't a particularly maritime country, but its oceanographers have long been among the very best: the first measurement of water salinity was taken in the Bay of Kiel in 1697 by Samuel Reyher, then a professor at the young Christian-Albrecht University. The term "plankton" was invented in 1887 by another professor, Victor Hensen, who sailed from Kiel to the North Atlantic and demonstrated that cold waters are more productive and richer in fish than tropical ones. In the 1960s, a geography professor named Günter Dietrich rebuilt Kiel's oceanographic institute (IfM*) so energetically that he died of a heart attack in 1972. In particular, Dietrich's team put a training program together (today we would call it a master's degree) that allowed students to do a few weeks of research in each department of the institute, which was resolutely interdisciplinary: physical oceanography, marine chemistry, planktology, fisheries biology, marine zoology, and marine microbiology all figured on the syllabus. Daniel received one of the best educations in the world in this branch of science. Twenty years later, I also went to Kiel, the way some people go to Mecca—I had dreamed of becoming an oceanographer since my earliest childhood.

---

* Institut für Meereskunde (Oceanographic Institute), also sometimes affectionately called Institut für Märchenkunde (Institute for the Study of Children's Stories).

I walked into the IfM library and kissed the ground; there, in the gray light of the Baltic winter, were thousands of volumes dedicated to the study of the oceans. Ecstatic, I inhaled the contents of my lecture classes, gorged myself on journals and books in four languages (including the fearsome *Allgemeine Meereskunde**), and muddled my way through the lab work, still lacking in practical experience. After fifteen years spent surviving the hostile and anxiety-provoking French education system, I began a second life in which work and study were no longer punishments to be borne.

But in the 1970s, Kiel wasn't just a university town. It was also a hotbed of industrial activity that revolved around huge shipyards, which employed over twenty thousand people. This "city within a city" was the center of intense labor negotiations and protests in reaction to restructuring, buy-ups, and waves of layoffs that have continued uninterrupted to this day. The shipyard has also accumulated its share of political and financial scandals, including the secret sale of military submarines to South Africa during apartheid in the 1980s and arms contracts with Saudi Arabia. Between these episodes, however, the site prospered, and a whole generation of immigrant workers, mostly recruited in Turkey, settled near the docks. Daniel put down anchor in that same working-class neighborhood—university housing was rare and mostly controlled by student unions who were better known for their fencing contests than their social policies.

Daniel went back and forth between these two worlds, from working-class tenements to the beautiful neighborhood that hosted the historic university buildings near the parliament of Schleswig-Holstein. He was also a social animal who knew how to make friends and form strong intellectual relationships, even in Northern Germany, a place hardly known for human warmth. Cornelia Nauen, who met Daniel on the second day of classes in autumn of 1969, tells me about this character trait as well as many other important details of his scientific career. Cornelia, then a dark-haired young woman of nineteen, had fled the

---

* "General oceanography"—a huge volume, well researched but heavy reading.

Ruhr ("a concrete desert") and a stiflingly conservative home. Daniel's seriousness and maturity impressed her, and he appreciated her in turn for her absolute loyalty, rigor, and industrious good humor. It was the beginning of a lifelong friendship during which they would share a scientific and political mission. Cornelia also studied at IfM, before doing her thesis on coastal ecological processes in the Baltic Sea. After that came six years of working for the FAO* in Rome on international marine biodiversity management and the fight against pollution. The FAO jubilantly handed out tenure to most of her male colleagues and "forgot" all about Cornelia, despite her creativity and effectiveness. In 1985, that venerable organization lost a high-powered scientist when Cornelia left to join the European Commission in Brussels, where she spent the next two decades developing partnerships with the Global South—concerning fisheries, of course. Since 2010, Cornelia has been the president of Mundus maris, an NGO that fights to protect the oceans and the people who depend on them using all of the scientific and artistic means at its disposal. Even after several hours of interviews, I wouldn't learn much more about her—a kind woman who obviously enjoys talking about the oceans, Daniel, and current events, she carefully avoids mentioning herself or her political engagements, however fascinating.

Indeed, during their time at the institute, Daniel, Cornelia, and several other students shared more than a passion for marine environments: they were also active in the Marxistische Studentenbund Spartakus, an organization close to the German Communist Party (DKP). This movement, following in the footsteps of Rosa Luxemburg and Karl Liebknecht, was the result of the student protests of the early 1970s. In Kiel, the Communist Revolution was no recent phenomenon. In fact, in October 1918, Kaiser Wilhelm II's navy had its very own "Mutiny on the Potemkin." The German sailors refused to go to sea for a final face-off with the British Royal Navy. Workers in the naval yards rallied to their cause and the insurrection spread from there, sweeping

* Food and Agriculture Organization of the United Nations.

across Germany like wildfire, forcing the abdication of the Kaiser and leading to the foundation of the First Weimar Republic, as well as the proclamation of a "Free Socialist Republic" by Karl Liebknecht. In Kiel, traces of this historic battle are more than discreet, almost hidden. In the 1990s, I heard about the great communist revolt of 1918 for the first time when a whole cache of hundred-year-old weapons and tracts was discovered, entirely by accident, in a tower during renovations at Kiel city hall. Daniel distributed modern versions of these documents down near the shipyard, and Marxist dialectics taught him how to conceptualize and defend his convictions with strength and perseverance, both orally and in writing. For Daniel, who has never been to a psychologist, "Marxism is also the negation of inner life; everything is dominated by exterior forces. It pushed me to ignore my own feelings and drown myself in work—that's always been my solution." For many of the leftist students in Kiel, the big question was whether or not to get involved with the RAF.* Unlike the Maoists (whose work Daniel had already seen in Paris), the Spartakists kept their distance, which turned out to be the right decision: the RAF's highly publicized career began with harmless pranks and ended in a bloodbath that led to a strong right-wing backlash in West Germany. Daniel set himself apart forever from "those idiots who plant bombs."

For the professors at the oceanographic institute, though, Daniel, Cornelia, and their friends were redder than red, and they teased their students by asking why they didn't simply head east, to live in the earthly paradise that had become the German Democratic Republic. Their professors still respected them, however—these politically active students were also the best in their class, the most brilliant, and the hardest workers. Daniel could often be seen wandering the halls of the institute, having long discussions with his classmates and professors about their coursework and current events. But what did he learn about the ocean during those years? Mainly that it is a giant physical, chemical,

---

* Rote Armee Fraktion, the "Red Army Faction," also known as the Baader-Meinhof Gang.

and biological machine of infinite complexity, whose many parts form a larger whole, a single dynamic system. In this way, the seasons, the winds, and the currents determine the growth of plankton, who then feed the fish, who then feed birds and marine mammals, and often humans as well. For more than a century, however, fisheries biology has developed somewhat apart from other branches of oceanography.* The reason for this difference is that there are very strong economic interests tied up in fisheries that make this area of study more "serious" and better funded than other marine disciplines, which in comparison, tend to look like nothing more than pastimes for humble naturalists.

FISHING: THE FIRST traces of this human activity date from tens of thousands of years ago. Today, fishing directly or indirectly employs 300 million people around the world and feeds 500 million people in developing countries, where it is often a major source of food. If we define it as gathering aquatic organisms in the most general sense, the earliest evidence of fishing in France can be found on the walls of Cosquer Cave (eighteen thousand years BP), near Cassis, in the form of drawings of harpooned monk seals. When it comes to harvesting ocean fish, some cod fisheries off the coast of Newfoundland are so consequential that they have actually changed the course of human history. For instance, they fed slaves in the Caribbean for centuries, creating a Creole culinary tradition based on a source of fish several thousands of miles away.[1]

Despite their paramount importance for entire civilizations in Asia, the Pacific Islands, and Europe, fish populations only recently became the subject of rigorous scientific study. In Northern Europe, the Norwegians were the first to begin to worry about large-scale fluctuations in cod stocks along their coast. Still the poor relative of the other Scandinavian countries, Norway was not yet one of the world's leading

---

* I use the term *oceanography* in the literal sense, meaning "the study of the oceans," which includes numerous subdisciplines. I am choosing to go against the French research community in physical oceanography, which has a vexing habit of claiming the term for its own exclusive use.

petroleum producers. The residents of its Lofoten Islands depended heavily on marine resources, and the government (under Swedish rule at the time) commissioned the first scientific studies of cod in the 1850s. Early progress was slow. It was not until the early twentieth century that Johan Hjort, director of the Institute of Marine Research in Bergen, understood that the fluctuations were determined by the rate of survival of cod larvae. Some years, many larvae survive: this leads to a "strong cohort." Other years, environmental conditions create a hecatomb: this is called a "weak cohort." Small fish become big fish, and strong cohorts lead to good fishing years while weak cohorts create scarcity for the Lofoten fishers.

Birth, growth, reproduction, and death (natural or caused by fishing): these are the fundamental processes based on which scientists attempt to calculate how many individuals can be harvested without causing the collapse of an entire fish population, the same way you prune a bonsai without killing it. The best biomathematicians went to work on the problem and developed increasingly complex statistical models that would be steadily improved by long-bearded European men, mainly during the first half of the twentieth century. The second half of the century was then devoted to studying different variations and perfecting existing models. Daniel began his career, therefore, in a field that already had a strong formal basis, on a stage dominated by some very impressive founding fathers.

The concept of overfishing was first defined in 1931 by a British scientist named Edward Russell, but in the 1970s, the world's fisheries were still in a state of euphoria, the catch having more than tripled since 1950 and coming ever closer to the symbolic record of 100 million metric tons* per year. At the time, each wild fish population was managed like a marine monoculture: the rules for herring were different from the ones for sprat, and neither had anything to do with cod. Daniel absorbed these doctrines—built around the study of some two dozen North Atlantic commercial species—without question. Two

---

* A metric ton is equal to 2,204 lb.

decades later, I did the same, though unlike Daniel, I developed a fascination for plankton, then marine birds—a specialization judged to be
much less profitable. Fisheries biology: that was the place to find work
as a marine ecologist.

And at the oceanographic institute in Kiel, fisheries biology was a
big deal, overseen by a major authority figure—Gotthilf Hempel.* This
formidable professor is the most influential German oceanographer of
the twentieth century. He began his career far from the water, however,
with a thesis on the physiology of grasshoppers at Heidelberg University in 1952. He was given a position in that still-lovely city, which
had escaped the bombs, but his chronic asthma pushed him toward
the coastal regions of Northern Germany. He moved there for good
in 1967 after several years of traveling around the world for UNESCO,
during which he built an impressive network of influential friends in
tropical countries. What followed was an exceptional career during
which he directed no fewer than six different research organizations,
founded the Alfred Wegener Institute for Polar and Marine Research,
and worked to get a 387-foot icebreaker, the *Polarstern*, built with the
staunch support of his friend, Chancellor Helmut Schmidt. Hempel
also published a colossal number of books and scientific reviews and
supervised more than sixty-six doctorates, easily making and destroying scientific careers along the way. Today, the word on the street is
that if he dies in time, the next German icebreaker will be called the
Gotthilf Hempel.

I attended several of Professor Hempel's classes in the 1990s, and
see him again in January of 2016 for an interview in Kiel. We agree to
meet at the oceanographic institute, now called GEOMAR. Hempel is
nearly eighty-seven years old, and his white hair and beard give him an
unearthly look. He pulls up in a shiny black Volkswagen Golf, accompanied by Cornelia Nauen, who has come up from Düsseldorf for the
occasion. I have reserved a quiet meeting room near the roof terrace,
but Hempel has his own agenda: he wants to sit in the library and has a

---

* *Gotthilf* literally means "God's help."

little less than two hours for me. He strolls off decisively in that direction. We find ourselves in a noisy and crowded setting where coffee is reverently served to the master and his guests. Quite a few acquaintances stop by, surprised to see us sitting at a table with a few books and two microphones. As a student, I would have fainted at the idea of speaking to Hempel, but I enjoy this interview. He is not an easy subject, too used to captaining his own ship. Still, I'm not interested in the official story, and I have come armed with a list of specific questions. Hempel finally gives in with good grace but finds my method much too disorganized; luckily, this test won't be graded. His mind, and his words, are razor-sharp as he paints me a picture of the young Pauly. "Daniel was formidably intelligent and he liked controversy; even when he knew he wasn't completely right, he drowned the other camp with his arguments. He was sometimes aggressive during those debates, but never mean. We were on opposite sides, politically speaking, and we often disagreed, but I respected him and the esteem was mutual."

"For Daniel," Cornelia Nauen adds, "being a Black man was a daily struggle. His visit to the United States really made an impression on him, and this consciousness, along with his Marxist ideas, was very important to him. He was already very loyal back then." In those days, Daniel was happy to talk about most things, but he remained discreet about his personal history; Hempel wouldn't learn the details of his past until thirty years later. "The hardest thing was the way other people looked at me," Daniel explains. "The first time I talked to Cornelia about my past, she started crying, so I didn't push it after that."

The rapport between Hempel and his students was very relaxed. Daniel never addressed him as "Professor Hempel," despite the German obsession with titles, and still calls him "Herr Hempel" today. The *Herr Professor* also had a mischievous sense of humor and soon procured himself a Little Red Book, memorizing whole sections in secret. In the students' office, a portrait of Chairman Mao took up a whole wall. During "strategic" discussions with his young colleagues, Hempel would seat himself beneath the portrait of the great leader and jokingly slip a few well-placed quotes into the conversation. Hempel's act delighted Daniel and Cornelia, who weren't fans of Maoist pomposity either.

Still penniless, Daniel worked at a nearby hospital. Though he received a scholarship from the Protestant Church* in his second year, he struggled financially during his first year in the program. He learned then that the oceanographic institute offered students one-hundred-hour contracts doing the most tedious laboratory chores. He offered his services and became assistant to one Wolf Arntz. Doctor Arntz, whom everyone calls Petz, was another child of the war, from Southern Germany. The son of a linguist, he spent his childhood wandering the banks of the Rhine, dreaming of endless oceans. At university, he initially studied English and Russian, thinking he would become an interpreter, but with mandatory military service hanging over his head, he decided to prolong his studies and go into marine biology, finally reaching the sea coast and its distant horizons.

When I meet him in Bremen in 2016, this epicurean octogenarian can look back on a long and agreeable career. Brought on as a public servant by the German agency for technical cooperation after his doctorate, he traveled the world and spent several years in Peru in the 1980s before becoming chair of marine biology at the Alfred Wegener Institute in Bremerhaven and participating in multiple expeditions to Antarctica. Everyone likes Petz. A bear of a man,† he enjoys making little delicacies for his friends and has even cooked on board the ice-breaker *Polarstern*, whipping up sumptuous meals for over a hundred people. In Bremen, he takes me to one of the best restaurants in the city, situated along the Weser River, and we chat happily about a young upstart named Daniel Pauly. "For Daniel, of course, I was a horrible reactionary, but he told me that at least I wasn't racist," Petz explains over his second cup of coffee. "He was obviously well ahead of everyone else and couldn't be bothered with details."

---

* "How a commie like Daniel managed to get a scholarship from the Church, I'll never know," comments Gotthilf Hempel.
† In fact, "Petz" is a familiar name for a bear in German.

Yet in the early 1970s, the task that Petz assigned to Daniel was an extremely meticulous one: sorting through mud. In fact, Petz was a benthologist, specializing in everything that happens on the ocean floor. Out at sea, much of his work involved hauling up big globs of sediment from the depths using an articulated mechanical digger whose jaws bite into the ground, then close around their contents; the whole thing is hauled up to the surface and emptied on board. This slimy stuff, which contains a number of tiny organisms, was then mixed with formalin and stored in jars labeled with the date and the location where it was collected. Such jars filled row upon row of shelves in the basement of the oceanographic institute, like green beans in a farmhouse pantry. Often years later, research assistants would open them to sort and identify the little creatures, whose number and diversity is an indication of the health of the marine ecosystem in question: when there are only a few worms, as is the case in certain "dead zones" in the Baltic, there is cause for legitimate concern.

When Daniel was assigned to mud sorting duty, the method used was very primitive: aliquots* were taken, then placed in petri dishes. After a quick rinse, the student would place the mixture of sediment, formalin, and preserved critters under a dissecting microscope to be identified and listed. This technique had several serious drawbacks. Students and researchers were exposed to formalin fumes, which are highly toxic. No one was worried about it at the time, but we soon learned that formalin is carcinogenic and that exposure over long periods of time can lead to potentially deadly allergic reactions. Additionally, the sediment swirling around in the mixture of water and formalin prevents the technician from seeing clearly, so sorting through the samples is terribly tedious. The whole process quickly frustrated Daniel, who began looking for a better way.

Prepared for this moment by years of working in industry, he decided to invent an automatic sorting system, putting together a

---

* A small quantity that is taken out of the whole, of which it is supposedly representative.

"very badly made but functional" prototype in his kitchen. Daniel's technique uses elutriation, the separation of particles based on their density: the mud sample is placed in a funnel, the bottom of which is covered with a screen. The screen is transpierced by a pipe, up through which a continuous stream of water is run, causing the sample to bubble. The little marine organisms, less dense than the rest, float to the top and overflow out of the funnel. They then run down the outside and are gathered in a filter attached to the base. The oceanographic institute had a large workshop and a team of friendly technicians who quickly transformed Daniel's invention into a beautiful machine that allowed samples to be separated five to ten times faster than before. "Daniel came up against a systemic problem and he found a technical solution," says Cornelia Nauen. "Since then, he's always taken the same approach—just with bigger and bigger problems." Daniel agrees: "My biggest advances have been mainly methodological." Characteristically, his German university had asked him to resolve a technical problem with his own hands rather than throwing himself headfirst into theoretical questions, as is often the case in the French system.

The mud-separating machine ran like a dream, and Daniel never completed his one-hundred-hour contract. Petz was magnanimous, and once Daniel reached his sample quota, he was allowed to seek new horizons. Hempel had kept well abreast of the affair, however, and asked Daniel to write an article describing his new machine; the student hopped to it and wrote a short paper, two pages in perfect German that would appear in an obscure review in 1973 after only a few minor corrections from his mentors. This was Daniel Pauly's first publication—two humble pages about mud.

But this first attempt was hardly trivial, as it allowed Daniel to go beyond his lectures and other coursework and start working in the laboratories of the oceanographic institute. At the time, Petz Arntz's and Gotthilf Hempel's labs were located on the docks, near the fish market. In the early seventies, German fisheries were still active. From Iceland to Greenland, they caught huge quantities of rockfish and pollock, while the coastal fisheries of the North and Baltic Seas provided heaps of cod, plaice, and flounder. The docks were a lively place, and from behind their laboratory window, Petz and his students engaged in some amateur sociology. The system worked like a gold rush: the fishermen would raid the northern waters and return with their pockets full of cash. As soon as their feet touched solid ground, they would run after any woman who chanced herself on the docks, including young students, whom they'd try to trap inside telephone booths. Since that hunt wasn't very successful, they would call a taxi and go to Hamburg, where they could *really* live it up. A week later, the ship's officers would follow them with a minibus to retrieve the human wreckage, and the now-penniless sailors would have no choice but to go back to hauling nets out of the North Atlantic.

Forty-five years later on a freezing January day, when I visit the area where the fish markets used to be, the only sound is the crunch of my boots in the snow. The magnificent red brick buildings are still there alongside statues of heroic sailors, but there is not a single fishing boat moored along the docks. The depots have been transformed into facilities for a start-up that makes electronics, mainly for the oceanographic institute, which has itself tripled in size thanks to new construction. Wandering through its cavernous lobby, I catch a glimpse of some comfortable-looking offices. The institute's annual budget of 90 million euros is also quite comfortable and funds the work of about forty professors. Of those, thirty-one are men—thirty Germans and one Pakistani, who I'm told grew up in Germany. In the twenty-first century, German marine science is still mostly a white man's game. The research interests of the venerable institution have, on the other hand, changed a great deal since Daniel's time here, and even since my time in the 1990s: geological and mining interests are prominent, as is the study of the ocean's great biogeochemical cycles. The climate and its impact are, of course, taken into consideration, but ecological questions are considered marginal, especially where vertebrates such as fish are concerned. Certainly, at a time when the oceans are being emptied of their birds and fish, it seems wise to shift one's focus to plankton and other tiny sea creatures.

DANIEL'S FIRST
AFRICAN EXPERIENCE

I N 1971, DANIEL visited Africa for the first time. He had been pestering Hempel to let him go since his second year, so the *Herr Professor* activated his extensive network of contacts and, after multiple messages to old coworkers at UNESCO, a partnership was finally set up with Ghana. Daniel received a plane ticket, a case of equipment, and some cash. His instructions were strikingly vague: have a look at the fisheries in the Sakumo Lagoon, near Accra, and collect data for the master's thesis he was to write at the end of his four-year program.

Daniel's wanderings in no way matched up with the university calendar, but no one seemed worried about him leaving to spend six months in West Africa. So, he discovered Ghana: the pioneering country of African independence, Pan-Africanism, and anti-imperialism, thanks to its leader Kwame Nkrumah. But Nkrumah had been overthrown by a military coup in 1966 (with some less-than-discreet help from the CIA), and he died soon afterward in Romania, never setting foot in his homeland again. In the early seventies, governments came and went, and corruption settled in, along with a serious economic slowdown. During his travels, Daniel wondered at the remains of an alliance with the USSR: dead tractors abandoned in the middle of fields and huge trawlers, ill-suited for coastal fishing, rusting in the port of Tema. It was in this seaside town that he would set up shop, at the fisheries research center near Sakumo Lagoon. The labs were on the ground floor, and Daniel shared the apartment upstairs with

Rodney, an American scientist employed by the FAO. People would come through Rodney's at all hours, day and night. Most of them were there to supply him with ganja, alcohol, and the semiprecious stones that he bought and sold to finance his drug habit. Rodney didn't hesitate to wander around slouched and disheveled in front of his religious and sharply dressed Ghanaian colleagues. Though Daniel didn't mind his roommate, he had to insist on several occasions that he really *had* come to work and that, no, he was still not interested in narcotics.

Jumping into his research with the relentless dedication that he has since become known for, Daniel devoured the center's stock of scientific literature, focusing on his reading every other day. The rest of the time, he wandered the lagoon, which was only about twenty inches deep—no need for a research vessel; a pair of strong legs would do just fine. These forays got him his first sunburn, an injury that confirmed his European identity, dispelling any fantasies he had about being able to assimilate in Africa. In his free time, Daniel met girls in town or wrote reports for old Herr Hempel on his portable Olivetti. In his letters, he describes the key features of his research. First, he showed that the salinity and temperature of the lagoon varied greatly depending on rainfall and the exchange of water with the sea coast and the Gulf of Guinea. This changeable environment was particularly determinant for the flora and fauna, and Daniel's tally of worms, shellfish, crustaceans, and fish painted a picture of a simplified ecosystem. He drew a schematic like the ones he had seen in ecology studies and soon, the lagoon came alive for him: "When I arrived at Sakumo, I didn't see anything, just a flat stretch of water. After a few months, I could see all the species that were present, their interactions, and the role of the fisheries—it was fascinating."

The fisheries in question were artisanal, centered on tilapia, the pretty tropical cousins of the perch, common in West Africa and also in aquaculture, so much so that they have established wild populations in many other parts of the world, including Florida. Daniel took around 1,600 samples, describing the tilapia's diet in detail and measuring their energy requirements, among other things. He also looked at how the local fisheries operated. The system functioned because basic rules were agreed upon by the dozen or so fishers who used the scant square kilometer of the lagoon: fishing was allowed six days a week, eight months out of the year. Time worked each day was around three or four hours, during which each fisher caught two to four kilos (between four and a half and nine pounds) of fish. Part of the catch would be sold for a dollar along the road between Accra and Tema; the rest would feed the fishers' families. Though he didn't know it at the time, Daniel was looking at an idyll of artisanal and subsistence fishing that would soon be steamrolled in the name of industrial profitability. He was destined to champion this kind of artisanal fishing, but that came much later. At the time, he pushed for development, which was thought to be both necessary and advantageous. His research report includes all kinds of proposals for how to "develop" fishing in the lagoon, via extensive aquaculture with tilapia or gray mullet, for example.

Ultimately, Daniel turned his master's thesis into four publications, which would be exceptional even today. Particularly because an ordinary student would have probably focused their work on only one aspect of tilapia biology, such as reproduction, and ignored the big picture. Choosing to go against this reductionist approach, Daniel produced an environmental and sociological X-ray of the whole lagoon and its fisheries.

During his stay, the fisheries research center in Tema held an international conference on the sardinella, a small fish that is abundantly available and ecologically essential in northwest Africa. Eight researchers from West Africa were invited, half of them from former British colonies, the other half from French ones. All the anglophones were Black Africans trained by the British; all the francophones white Frenchmen from ORSTOM.* "They looked like pawns on a chessboard," Daniel recalls, "and no one said anything—it would have been like mentioning rope in a hanged man's house." But of course, Daniel couldn't keep his "big mouth" shut: "Why don't you have local partners who are trained

---

* Office de la recherche scientifique et technique outre-mer.

to replace you?" he asked. The researchers from ORSTOM weren't racist, but training African colleagues simply was not part of their mission at the time—a policy the organization didn't change until "they had been thrown out of several Black African countries and Madagascar." Only then was the French Institut de recherche pour le développement (IRD) created specifically to set up partnerships with countries in the Global South. In 1970s Ghana, however, Daniel's words hit a little too close to home, earning him the dislike—almost hatred—of an entire generation of French overseas researchers. Remembering the episode, Daniel quotes Jean de La Fontaine:

> According as you're feeble, or have might,
> High courts condemn you to be black or white.[1]

At the end of 1971, before returning to Germany, Daniel spent two weeks traveling around Ghana. Heading north, he visited Akosombo Dam, which had only been operating since 1965; dead trees could still be seen sticking out of Lake Volta. He crossed this giant stretch of water on a strange platform-like boat—even the bus was loaded aboard. Later, near the edge of the Sahel, he spent some time in a small farming village among the goats and chickens. It occurred to Daniel that his African ancestors might have lived like this, and if so, they were "very respectable ancestors indeed." He returned home changed by what he saw in that dry country. "Before that, I didn't know what it meant to be African. You could say it was a pilgrimage." As in the United States and France a few years before, Daniel had reached his journey's end. He returned to Kiel at peace, made all his friends sick with fermented corn brought back from Accra, and wore brightly colored clothes for a while, though they soon shrank in the wash. He returned to his classes as well and, at night, processed the many jars of samples he had brought back from Sakumo.

In the late 1990s, Daniel returned to Ghana and saw the lagoon once again, but it was different, surrounded by encroaching urban development. The fish were even smaller than before, and Daniel wondered if Sakumo might simply disappear someday. Since 1992, the area has been protected by the Ramsar Convention,* but this designation does not seem to have stopped the forward march of urbanization.

Back in Kiel, the Protestant Church's merit scholarship† allowed Daniel to concentrate on his studies without worrying about money, but the award also came with certain obligations. The first summer, he had to go door to door raising money. The following year, however, the deserving student was allowed to take an intensive language class. During the summer of 1972, Daniel chose to learn Russian in Bochum, surrounded by future West German diplomats. German universities have excellent language-immersion programs. I experienced them for myself in Heidelberg in 1989, where I arrived without speaking a word of German and, in ten months, was rocketed to a level high enough to take university exams. With around fifty Russian words and expressions to memorize each day, Daniel advanced rapidly and soon began watching the great classics of Soviet cinema that the university screened (without subtitles) in the evenings. In the summer of 1973, after a "refresher course," Daniel participated in a three-week tour of the Soviet Union, via Moscow and Leningrad, where he admired the Hermitage. He found himself in Kiev on September 11, 1973, when Salvador Allende was overthrown and assassinated, and Chile hidden from the world behind a wall of secrecy. "I saw people crying in the streets of Kiev—you can't tell me it was just propaganda." Daniel developed a deep sympathy for the scientists of the Eastern Bloc, who, despite their monumental achievements, remained largely unknown to the international community because they did not publish in English. He studied the great masters of Russian and Ukrainian fisheries science: Fedor Baranov, Lev Berg, Viktor Sergeevich Ivlev, and Peter Moiseev; and the writings of Georgii G. Winberg and Georgiĭ Evgen'evich Shul'man on the physiology of fish would have a decisive impact on his own scientific thought.

---

* The goal of the Ramsar Convention, named for the Iranian town where it was established by UNESCO in 1971, is to protect wetlands around the world.
† Evangelisches Studienwerk Villigst, evstudienwerk.de.

In 1973, Daniel took a series of long-distance trips, the first time he had taken so many in a row, visiting one destination after another with an appetite for travel that would only grow over the next few decades. In addition to his stay in the Soviet Union, he spent a whole semester in the United States and participated in an oceanographic research mission up the coast of Canada, going all the way to Greenland. These trips more or less fit into his curriculum, and nobody in Kiel seemed worried about his growing number of absences—he already had all the credits he needed to graduate, with brilliant exam grades to boot.

In the US, he went on an exploratory mission with the goal of writing a summary report on catfish farming. Daniel visited numerous aquaculture sites in the southern half of the country and wrote a paper that would make him, for a while at least, the German specialist on the subject.[2] He also contacted Lester Aronson, a specialist in tilapia behavior at the American Museum of Natural History in New York. Daniel hoped to learn more about the fish he had studied in Ghana, but Aronson was reluctant to meet, having recently suffered a public image problem after being accused of vivisecting neighborhood cats. Daniel managed to convince him, however. "The old man was expecting to meet a 'German researcher,'" Daniel recalls. "He was very surprised when I showed up."

His third trip across the Atlantic was also an opportunity to re-immerse himself in African American culture with a visit to his cousin Robert Whitfield (Ed's brother), who was going through a personal crisis at the time and living in Bedford-Stuyvesant,* a majority Black neighborhood in Brooklyn. "I saw some horrible things, mainly having to do with drugs and poverty, a real descent into hell, right through the middle of it. There were social services, but they acted like an occupying army, doing inspections on these poor families incessantly, even at night. I would have written my thesis on it," Daniel concludes, "except that I was supposed to be a fisheries scientist." Horrified by what he'd seen in Brooklyn, Daniel returned to New Rochelle to stay with Uncle

---

* Spike Lee filmed *Do the Right Thing* there in 1989.

Carl, who invited him on a cross-country trip. The duo set off for California, where they only stopped at Daniel's father's house long enough to borrow a car so that Carl, a real gambling man, could drive Daniel down to Vegas.

That year, 1973, Daniel also participated in his first research trip on the high seas. For students at the institute, this was a rite of passage, a highly anticipated moment for everyone who chose to study the oceans, dreaming of the works of Joseph Conrad and yearning for long sea voyages as they studied in some gloomy corner of the university dormitory. Daniel, for his part, headed for St. John's in Newfoundland to go on a five-week mission aboard the *Walther Herwig*, a huge trawler that had been converted into a research vessel for the German Federal Institute of Fisheries.* Daniel left New York by bus and saw Canada for the first time. "The first thing I did was go buy a hockey stick, a nice, solid hockey stick that I still have—I'd dreamed about having one since I was a kid." He had made the trip with Cornelia Nauen's boyfriend, who would be his crewmate on board. The two young men got along, but just barely—Daniel and Cornelia's friendship created some jealousy, likely exacerbated by the brief liaison Daniel had pursued with an Iraqi student after their arrival in New York. The pair would soon learn to stand shoulder to shoulder, however.

The *Walther Herwig*'s mission involved surveying populations of groundfish between Newfoundland and Greenland, via Labrador and the southern coast of Baffin Island. Many a fishing fleet coveted the region's cod, rockfish, and halibut. "St. John's Harbor in Newfoundland was full of boats from all over the place: Portugal, Poland, East Germany, etc." Daniel witnessed a trawler at work on an industrial scale for the first time: "The *Walther Herwig* scraped the bottom at a depth of 500 meters (1,640 feet) and hauled up a mixture of rocks, cod, and deep-ocean fish. Some of the glacial erratics† we pulled out of the

---

* Bundesforschungsanstalt für Fischerei.
† A glacial erratic is a piece of rock that is trapped inside an iceberg, then dropped into the ocean when the iceberg melts.

water were as big as Volkswagens. We measured fish, removed otoliths with knife points, but the data we collected was absolutely useless
because the fisheries weren't regulated. There were no quotas, it was
a free-for-all, a gold rush for cod." .

After three or four days, Daniel managed to get over his seasickness.
"I always get seasick, sometimes while we're still at port—diesel fumes
make me nauseous—but it usually goes away after a while. When I'm
leaning over the toilet, I always think about Darwin. The poor man
was sick for the whole voyage of the *Beagle*." The atmosphere on board
was tense, and the two students found themselves isolated from the
sailors and the institute's research assistants, "old farts who had spent
decades measuring fish on the Hamburg docks."

Remembering his shipmates, Daniel becomes vehement: "They
didn't like us because we were students; the older research assistant
grumbled that students don't work, they just sleep around and dis-
tribute leftist pamphlets." The crew complained constantly, but the
roughnecks kept saying that once they made it to Nuuk, it would be
heaven on earth because "there's girls and alcohol." "We wondered
what they were talking about because, in our minds, Nuuk* wasn't
exactly Las Vegas," says Daniel.

After a few weeks of trawling, the *Walther Herwig* entered the Davis
Strait on the border between Canada and Greenland, which was still
a Danish colony. Playing to the crowd, one of the research assistants
claimed to be able to determine the sex of rockfish by inspecting their
cloaca.† In fact, males and females of the species are perfectly identical,
and the scientists on board expressed skepticism, especially Per Kanneworff, a Dane who, Daniel recounts, "asked, in perfect Scandinavian
style, 'May I test your method?' And the idiot accepted." A hundred
or so rockfish were randomly chosen, then inspected by the research
assistant and sorted into two buckets, supposedly containing the males
and females, respectively. Per the Dane then cut open the beautiful

---

* The capital of Greenland, then called Godthåb.
† Anal and reproductive orifice.

multicolored fish with a big knife and inspected their gonads. "Of course, you can't sex individuals by looking at the cloaca—I've never seen such an embarrassing situation," Daniel recalls. "The guy from the fisheries institute had twenty years of junk data on the sex of rockfish that he couldn't use anymore. That ridiculous challenge just made the atmosphere even worse."

The *Walther Herwig* finally stopped in Nuuk for a few hours and Per Kanneworff, who lived there with his family, invited the two students to his house, a pretty wooden building painted in bright colors. They spent the evening chatting over tea, then visited the nearby town and its fish processing plant. "I saw a kind of poverty that went totally against Denmark's official line," Daniel tells me. "I was shocked to see the Greenlandic equivalent of an inner city, complete with public housing and piles of trash barely hidden under the snow. I saw a local man standing on the side of the road with a halibut that he was trying to sell. Three hours later, he was still there; it made me sick to think that on the *Walther Herwig*, we had thrown hundreds of tons of edible fish overboard." Per the Dane stayed ashore, but the two students returned to the *Walther Herwig* around midnight, as they were supposed to head back to Germany early the following morning. About seven or eight hours later, they emerged from their rooms after a good night's sleep, happy to be roused by the sound of the motor and not the siren that had tortured them every morning at 5:00 AM during the previous weeks.

To their surprise, they were not heading southeast toward Europe but seemed to be returning to Nuuk. "No one on the whole ship would speak to us!" Daniel remembers. In fact, the *Walther Herwig* soon made port again in Nuuk, but only for an hour, after which it put out to sea again, heading toward Bremerhaven. "It was only after two or three days that an officer finally told us what had happened," Daniel says. That unhappy night in Nuuk, some Greenlandic girls had come aboard, drawn in by the promise of a heavy dose of alcohol. "They made them drink, then they raped them." One of the young women passed out in a corner of the ship, and they found her half-dead the following morning after the boat had already left port—hence the hasty change of direction. "They wanted to cover it all up—it made me so angry. They had always excluded us because we were students, leftists, good-for-nothings, but they turned out to be rapists, criminals."

Daniel does not remember anything about the crossing from Greenland to Germany, only their arrival in Bremerhaven. "Their wives were all there, their dear husbands were finally home and everyone was happy. They gave them tons of fish that they had stored away during the trip." Daniel went straight to Hempel and spilled the whole story, telling him about how the crew and research assistants of the *Walther Herwig* were apparently committing rape every time they passed through Nuuk. Daniel threatened to contact the press, but Hempel, who had a great deal of influence with the German research flotilla, assured him that no Greenlandic girl would ever set foot aboard the trawler of shame again. The pack of criminals went unpunished, however, and the episode continues to haunt Daniel: "It really disturbed me, it meant that you didn't have to go far to find real brutality." To this day, he still feels a strong aversion to the machismo and violence of certain "maritime professionals."

DEVELOPMENT AID
IN INDONESIA

A MAN IN RAGS stands waist deep in water, squinting intensely into the ocean. In one hand, he holds a piece of rope which, spread into an arc on the surface of the water, helps calm the wavelets, allowing his experienced eye to pick out the ocean's minuscule inhabitants. With his other hand, he skillfully collects tiny fish larvae only a few millimeters long with a spoon. They are not all desirable—most will be left in little piles along the pebble beach. Only those of the milkfish (*Chanos chanos*) will be carefully collected and stored. At the end of his workday, the man will sell them to an aquaculture business that raises the fish in pools. The adult milkfish will later end their days in one of the myriad kitchens of Southeast Asia, on the grill or in a soup.

It is June of 1975, on the north coast of Java, near Tegal, about 190 miles east of Jakarta. Daniel has just arrived from Germany as an employee of the German Agency for Technical Cooperation (GTZ) to help develop Indonesia's fisheries. Seeing the man collecting fish larvae, Daniel called out to him, "How much do you make in a day?" The man named a sum equal to a few tenths of a dollar. "That's not very much," Daniel said. "Yes," the fisherman answered, "especially because I have to give half of it to the owner of the rope."

Daniel would never forget this archetype of the "small-scale fisherman." The encounter led him to question the relevance of his development mission in Indonesia. "We were there to establish

65

a flotilla of trawlers," he recalls. Those bottom-scraping ships were already at work in the Gulf of Thailand and in Malaysia, cleaning out areas where fish had remained relatively abundant into the 1950s. The profits enriched a handful of fishing bosses while legions of villagers in the coastal areas got by on their leftovers. The GTZ's approach was actually counterproductive, but Daniel decided not to bring it up with them—and for good reason. After finishing his master's thesis, brilliantly defended in Kiel in February 1974, he had been lucky to get this job. The GTZ had smelled a potential leftist troublemaker, and Herr Hempel had had to throw the full weight of his influence behind Daniel's application to ensure that his protégé landed his first contract. Daniel did not have that kind of luck with his application at ORSTOM, the French development agency, at the end of 1974. The job, focusing on tilapia fisheries in Ivory Coast, was a perfect match for him given his experience in Ghana, but they turned him down. Two decades later, he learned from Jacques Daget, the eminent ichthyologist who was in charge of the selection committee at the time, that he had been their first choice among the thirty applicants. But a call from the Ministry of the Interior changed their minds—no draft dodgers were allowed in the French overseas civil service, and in the end, ORSTOM recruited a more docile biologist who would, in fact, not make a single wave during his entire career.

In Germany, the GTZ initially told Daniel that he would be working on a collaborative research project in East Africa. He spent several months in Bad Honnef, near Bonn, where a personal tutor from Zanzibar drowned him in Swahili while he learned about the rights and duties of a German development expert. His curriculum also included several weeks of training at the venerable Senckenberg Museum in Frankfurt, where he brushed up on fish classification. The course content proved to be extremely vast: planet Earth is home to several tens of thousands of species, most of which live in the coastal waters of tropical countries. Daniel was excited—a large project was taking shape in Tanzania—but after several long months of waiting, everything went sideways and he ended up being sent to Indonesia, where his Swahili would be of little use. In accordance with GTZ policy, he was assigned an Indonesian partner. Purwito Martosubroto was exactly Daniel's age and had just finished his own master's degree at the University of Miami in Florida. He was also paid about a tenth as much as his European colleague. Purwito was too good a man to let this detail bother him, though, and a tragicomic experience soon turned the two of them into fast friends.

During their first field mission in June of 1975, their driver was going a little too fast down a muddy dirt road. "I saw it coming and rolled myself into a ball," Daniel remembers. Their Jeep smashed into a tree—the driver's face was badly mangled and Daniel was "covered in cuts." Purwito, only slightly hurt, flagged down a taxi to take the driver to the hospital. Daniel waited on the side of the road, dazed. The big container filled with fish and formalin that they had been transporting had spilled all over him, and he was soaked in the smelly, corrosive liquid. A pretty young woman emerged from a nearby house to give him a glass of milk and a towel; he thanked her and did his best to wipe the stuff off. Another beautiful woman appeared and asked sweetly if he was all right, then another, then another. Daniel finally realized that he was sitting on the doorstep of a country brothel when a truck driver emerged, still zipping up his pants. The madam came out of her room to ask Daniel, still dazed and bleeding, if he would like to come sit in the shade for a while. When Purwito came back to get him, he made Daniel promise never to reveal the exact location of their accident to his wife.

Purwito quickly guessed the political convictions of his unusual European colleague and did his best to ensure that Daniel's leftist sentiments remained hidden. In 1970s Indonesia, being a communist or even a trade unionist meant death. Beginning in 1965, Suharto, a genocidal dictator, had organized an incredibly violent crackdown that caused the death or disappearance of one million people in just a few months. "My girlfriend told me that there were a lot of bodies floating in Jakarta

Bay—people stopped eating the fish," Daniel says. Ten years later, suspicion, public denunciations, and corruption still dominated Indonesian society. "Any undertaking was an infernal triangle consisting of a military officer, a straw man, and a Chinese businessman," comments Daniel. He lived in fear for the first time—it didn't take much to be denounced, tortured, assassinated, or deported to a prison camp on Buru in the Maluku Islands, over a thousand miles from Jakarta.

Though it wasn't easy, I managed to contact Purwito Martosubroto, who is now retired and living in Jakarta. During our phone conversation, politely and in a very calm voice, he confirms that during Daniel's stay, Suharto's martial law applied to everyone, including foreign diplomats.

On Cendana Street in Jakarta, Daniel lodged among Indonesia's elite: during his first year there, Suharto was one of his closest neighbors. Daniel lived with a grande dame of the former ruling class, by then an impoverished lady who discreetly received visitors from her past life and spent her time reminiscing about the old days. Daniel found this atmosphere oppressive and soon moved to Semarang—a calmer and more tolerant city—with Elizabeth Kambey, his Indonesian girlfriend, where they avoided gossip by pretending to be a married couple. The house next door was under construction at the time and a few indigent laborers scratched at the dirt with dull tools. On his way out the door one day, Daniel forgot himself and started whistling "The Internationale."* One of the workers stood straight up as though struck by lightning. Brandishing his shovel and crossing it with a stick he had grabbed from his neighbor, he cried, *"Bagus!"†* with a toothless smile. Then, in the same moment, he went right back to work in the trench. The poor wretch never looked Daniel in the face again.

Our apprentice researcher also spent six weeks at sea aboard the *Mutiara 4*, a 115-register-ton‡ research trawler, slowly crossing the Java Sea and South China Sea toward Singapore. Eighty one-hour-long

---

* A well-known left-wing anthem written in 1871 by Eugène Pottier. (TN)
† "Good!"
‡ Gross register tonnage (GRT): formerly the international unit used to measure a ship's internal volume; one register ton equals one hundred cubic feet. (TN)

tows* were carried out at regular intervals and the contents of the net dumped out on deck. "A typical tow would yield two hundred kilograms (440 pounds) of fish, and there were 150 different species, of which only 80 were known!" recounts Daniel. Biological diversity is at its height in the tropics. Daniel had read as much in Kiel and Frankfurt, but he had to see it to believe it. In the nineteenth century, that same incredible diversity had attracted Alfred Russel Wallace, co-inventor of the theory of evolution, to the region, where he eked out a living gathering bird and insect specimens for wealthy British collectors.

Over the years, Daniel has told the story of his expedition on that trawler in the Java Sea over and over again to anyone who will listen. It was the experience that launched his career in the tropics—a total shock, a wake-up call. As Cornelia Nauen puts it: "From there, he would have to reinvent fisheries biology."

Why? As we've noted, the management of fish stocks had, at this point, only been theorized by Northern researchers, and only for the cold waters of the North Atlantic. The number of fish species under consideration was limited, often to fifteen or so, making it possible to study each species in detail. More specifically, in order to manage a fishery, one must know the age of the fish, as well as their size and mass as a function of age. After that, in order to determine how much a fishery should catch, one must identify the natural mortality rate (that is, mortality caused by things other than fishing). For each species, this means years—decades, even—of work collecting samples at sea and analyzing them in a laboratory. To determine the age of fish caught, scientists since the beginning of the twentieth century had been relying on the work of Johannes Reibisch, yet another inventive scientist based at the University of Kiel.

Reibisch cut up and examined fish otoliths under a microscope, an extremely meticulous technique. Otoliths are calcium carbonate structures, like tiny teeth, that float freely in the inner ear of

---

* A "tow" refers to a single use of a trawl net, a process consisting of deploying the net, pulling it for an extended period, and retrieving it.

vertebrates—yes, you and I have otoliths in our vestibular systems too. Their position and movements are determined by gravity and they stimulate little hairs that line the inner ear, allowing us to keep our balance. If they don't work properly, we end up with a terrible feeling of seasickness called positional vertigo. I've had it—it's awful. Otoliths, which can be more than one inch long in certain fish, have a great aesthetic beauty, and for researchers they contain some fascinating information. In fact, once otoliths have been delicately sliced up and placed under a microscope, they reveal annual growth rings, like those of a tree. By carefully studying these cross-sections, researchers can determine the age of the fish they came from. Moreover, each species has distinctly shaped otoliths, forming a whole gallery of artistic creations. Even if these structures are all that remain of a bird's or a marine mammal's dinner, some quick forensic analysis can reveal the kind of fish they ate. I used this very technique to complete my thesis on cormorants. I examined tens of thousands of the tiny structures—I especially like the otolith of the cod for its rough, shell-like surface, and that of the sea bass, which has a unique protrusion that looks like the prow of a ship.

It is the Danes who have become the top otolith-slicing specialists in fisheries biology. They have made it into a real industry, aiming to control the global market for their particular skill. In the mid-twentieth century, this goal seemed attainable: development programs usually tried to copy-paste Western methods onto tropical ecosystems. However, as Daniel quickly realized looking at the hundreds of different kinds of fish hauled out of the water by his trawler in the Java Sea, patiently studying the growth of each species and then calculating how many the fisheries should catch would be impossible. In fact, the otoliths found in tropical fish do not have annual growth rings—and even if they did, the Northern European process is so time-consuming that the new trawlers financed by the Asian Development Bank would surely catch all the fish before scientists had time to learn their secrets.

Daniel began looking for a solution, but he wouldn't find it right away—he was frustrated, and his job proved more frustrating still. He had to tolerate a bad-tempered project manager—a shifty man who was a little too keen on the bottle. Daniel's scientific writings from his years in Indonesia are modest, initially limited to a few reports, then a single publication in 1980. It was not until twenty years later that Pauly and Martosubroto, still close friends, set things to rights by publishing an enormous tome on the fish of western Indonesia.

In the meantime, Daniel's contract went unrenewed, and the GTZ recruited a replacement behind his back. "Daniel returned to Kiel very suddenly," Hempel recalls. Exiled from Indonesia, Daniel suffered from more than unemployment—he and his girlfriend, Elizabeth, broke up for good when he left Jakarta. Back in Germany, Hempel received him coldly. The master was deeply dissatisfied with the work his protégé had done for the GTZ. Daniel explained the situation with his German project manager, and Hempel "took a detour" through Jakarta during his next globetrotting adventure to check his student's story. The project manager arrived drunk for their meeting in Hempel's hotel lobby. On his return to Kiel, the *Herr Professor* reassured Daniel and offered him a part-time position as a research assistant* so that he could start his doctoral thesis in January of 1977.

The length of a doctoral program in the sciences varies greatly according to the candidate and the country. Some go on for a decade or are never finished, but most are completed in three to four years. Daniel wrapped his thesis up in two and a half years, doing most of his research in 1977 and 1978.

When it came to directing theses, Hempel took a perfectly binary approach. Either the candidate was independent and creative, and Hempel offered only discreet and non-invasive support. Or they struggled and he chose a topic for them, drawing on the institute's seemingly infinite collection of jars brought back from past oceanographic expeditions. The victim would then spend several years with his or her nose

---

* This is how doctoral theses in Germany are usually financed. Everyone knows that PhD students are the ones who push research forward, but they are only paid to work part-time, despite the fact that they generally put in twice as many hours as their tenured colleagues.

stuck to an ocular lens, churning out a descriptive monograph on the state of plankton or fish larvae in some obscure part of the Atlantic.

Knowing Daniel, Hempel offered only a few general suggestions at the beginning of his PhD program, which Daniel resolutely ignored. After his Indonesian experience, he knew that he had to find a viable technical solution to the problem of managing tropical fisheries, and nothing less would do.

To accomplish this mission, he dove into the universe of Ludwig von Bertalanffy. This Hungarian theoretician was born at the beginning of the twentieth century into the Viennese bourgeoisie, surrounded by a star-studded family of churchmen, artists, intellectuals, and senior imperial officials. Initially educated by private tutors like all the sons of his social class, he went on to the universities of Innsbruck and Vienna to pursue a career in biology and philosophy. Going against the positivist approach* so in vogue at the time, he considered the universe to be a living system that exists independently of human societies.

During a long career in Europe, then in North America, with the support of his wife and close collaborator (who was, however, never listed as a coauthor on any of his work), he developed general systems theory,† which we now consider to be one of the most important conceptual leaps of the last century. Von Bertalanffy worked his whole life to bring the natural and social sciences together, but in the 1930s he also created, almost as an afterthought, an equation that describes the growth of living organisms. He defined growth over time as the result of a conflict between all the biological processes that allow an organism to live and grow, such as the synthesis of its molecules, cells, and tissues, and the degradation of these same molecules, cells, and

---

* Positivism: a system of thought that highlights scientific analysis and hard facts and believes in the benefits of knowledge. Even today, the idea that science and technological progress serve humanity and will resolve all its problems in the end is deeply rooted in our Western societies.

† General systems theory is that all systems, be they social, biological, or psychological, are organized in a way that is independent of their nature and the spatio-temporal scale at which they occur. This theory is used in cybernetics, psychology, and linguistics, among others, and is the object of transdisciplinary studies as well.

tissues. If the balance is negative, death will win out. If it is positive, the creature will be able to accumulate energy, grow, and reproduce. Von Bertalanffy's differential equation, almost as elegant as the one for special relativity, effectively describes the growth of many creatures, including fish. The British duo Raymond Beverton and Sidney Holt, giants in the field of fisheries biology, were not wrong to place von Bertalanffy's equation at the center of their seminal work, published in 1957, which laid down the foundations of the discipline. Analyzing the growth rate of fish is, in effect, an essential step for anyone who wants to manage a fishery. Small fish grow into big fish . . . and if they are caught before that point, then the fishery is not being managed rationally.

During his studies in Kiel, Daniel eagerly devoured Beverton and Holt's book, where he encountered von Bertalanffy's work for the first time. However, it "ran off his back like water off a duck." When he returned from Indonesia, though, he rediscovered the Viennese scientist, this time in the original German, in von Bertalanffy's 1951 book *Theoretische Biologie*.[1] Obsessed by the idea of being able to calculate the growth rate for the thousands of tropical fish he had encountered, Daniel realized that he would have to come up with a general model for the growth of *all* fish, without waiting for studious measurements based on cross-sectioned otoliths. To achieve this, he threw himself into a giant data-collection campaign, looking for all the information ever published on the size of fish as a function of their age, which allows the estimation of their growth rates. The word did not exist yet, but today we would call it a meta-analysis based on a review of published data whose goal was to determine some overarching principles of how the living world functions. Daniel took his inspiration from the Russian scientist Georgii Winberg, but his research project seemed totally eccentric compared to what his classmates were doing. In fact, the other students generally focused their doctoral research on a single species in one specific area and devoted much of their time to collecting and analyzing samples. Students love going on scientific expeditions aboard oceanographic research vessels, if possible in the more exotic regions of the globe—finally, a little adventure after all those years of hitting the books under the gray skies of Schleswig-Holstein.

Daniel, who had already roasted under the tropical sun, chose another battlefield: the library of the oceanographic institute. There was no internet and no databases—the information Daniel sought was scattered among the thousands of volumes stored in that venerable institution. The young researcher, who was "less a collector than an accumulator," spent most of 1977 and 1978 in the library. He consulted some works on site, but also borrowed boatloads of them—usually a dozen a day. The benevolent librarian, Frau Wollweber, tired of checking out all those books, eventually gave him carte blanche. After that, Daniel gathered his data as he pleased, creating one index card per fish species encountered in his reading. In the end, his collection contained 515 species from 978 different populations. Gathering such a dataset in the field would have taken several lifetimes, and Daniel, impressed by the results, never wasted time on fieldwork again, insisting, "It just doesn't go fast enough."

Among the populations included in his note cards, there were about a hundred that had been studied in sufficient detail to determine their natural mortality rates, a variable represented by the letter "M." The all-important "M"—holy grail of fisheries biology. Indeed, in order to determine how many fish can be sustainably caught by a fishery, you have to know how many die a "natural" death. In many tropical regions, Daniel's colleagues were starting from the assumption that all small fish grow and reproduce rapidly. They assumed, therefore, that all the populations renewed themselves yearly, and assigned "M" a very high value. Daniel didn't believe it for a second and saw no reason to expose tropical fisheries to such discriminatory simplification. Such assumptions are particularly dangerous because if analysts assume that populations are completely renewed each year, fisheries will be encouraged to treat the ocean like an open bar.

Daniel therefore constructed a statistical model derived from von Bertalanffy's equation that makes it possible to calculate M for any fish population based on the maximal size (or mass) and growth rate of the average individual. "I've never been very good at math, but I managed," admits Daniel, who relied on his vivid imagination to come up with "a sort of pseudo-code." Armed with a rapidly scribbled proposal, he went to see Gerd Woller of the physical oceanography department, who coded the necessary computational routines for him in FORTRAN.* The entire institute shared a single computer, which was connected to a number of terminals. The machine would allocate a fraction of a second of its calculation time to each operator, one after the other. Daniel punched in his data and a cloud of oblong dots appeared on the screen—these were the M values as a function of fish size and growth. Examining his results closely, Daniel noticed a striking pattern: the warm-water fish were clustered together on one side of the screen, the cold-water fish on the other. Not surprising, as scientists had long known that fish grow faster in warm water. Daniel added this factor to his statistical analysis and found a strong correlation between M, the maximal size/mass of fish, their growth rate, and the average temperature in their range.

Because Daniel had done such good work, Hempel tapped him to be part of the German delegation at the annual conference of the International Council for the Exploration of the Sea (ICES) which, in the spring of 1978, took place in Copenhagen, four hours north of Kiel by train. It was the biggest event of the year in fisheries biology, and the idea of presenting his research to the gods and demigods of the era—including the great Ray Beverton—terrified Daniel (though, as always, he hid it well). "I expected someone to say, 'This thing doesn't make any sense; it's missing this or that.' I thought they'd tear me apart," he admits. In the end, though, they thought his work was just fine.

Back in Kiel, Daniel completed his dataset and checked his analysis for a total of 175 species.[2] The resulting scientific article became his first big hit. "People either needed the equation or the database," he concludes. In fact, Daniel had very astutely published his entire database as a table in the article. Transparency and open access to raw data have

---

* A programming language.

been central features of his work ever since. The "Pauly equation," as it would soon be called, became popular because it allowed researchers to determine M using very simple parameters, including temperature. This earned Daniel a lot of kudos, but also some ridicule. The latter camp is best represented by his American colleague Ray Hilborn, who later criticized Daniel's method for being too low-tech and at odds with more complex, time-consuming methods commonly used in the Global North: "The Prophet Daniel is among those excluded. Some say that Daniel chose his exile, others say he was cast out. Whatever the reason, Daniel resides in the lower regions, the hot places. [He] must toil in infernal heat, deprived of holy catch-at-age data and armed only with a thermometer."[3]

The critiques were not always very sophisticated, though, and his publication soon found a privileged place in the scientific literature. Used by thousands of his colleagues, it is one of the top three most cited works by Daniel Pauly. Having made so much progress, Daniel could have done the next logical thing: write up his analysis in thesis form, bag his PhD, and then go looking for a job. For Daniel, though, his foray into the land of M was just a detour, and for his thesis, he wanted to go even further.

The von Bertalanffy equation actually predicts that organisms like fish will eventually reach a maximal size and stop growing, which is the case for all fish, from the guppy to the whale shark. However, good old Ludwig didn't specify the biological mechanism that underlies this phenomenon. Intense academic debates among his younger colleagues ensued—one scientist even suggested that they simply abandon von Bertalanffy's equation altogether. After carefully reading von Bertalanffy's texts in German, however, Daniel discovered that the master had left a clue behind: he mentions that the metabolism of living organisms is limited by "a biological surface," but doesn't take the time to explain which one. Daniel could see two possibilities: the intestines, which limit the absorption of the products of digestion, or the gills, which limit the supply of oxygen to the fish's body. Having cut up quite a few fish in Ghana and Indonesia, he decided that the surface of the intestine could not be the limiting factor, and turned away from that possibility. Around the same time, a Californian physiologist with a passion for birds, none other than Jared Diamond, took this constraint more seriously and showed that it controlled the life of hummingbirds: their intestines' capacity to digest nectar from flowers determines their energy gain and therefore their daily flight time. Pursuing his gill theory, Daniel dove back into his database and promptly showed that, in fact, the greater the surface of a fish's gills, the stronger its growth.

Already familiar with the effects of temperature on metabolism, he also postulated that the higher the ocean's temperature, the more oxygen the fish would need, putting more pressure on their gills. In order to explain the impact of temperature, he visualized the following phenomenon: the proteins that make up living bodies function properly because of their unique three-dimensional structures. In fact, many biochemical reactions are initiated or controlled by lock-and-key-type mechanisms, where a protein fits into a particular receptor. However, above a certain temperature, proteins collapse like an overcooked soufflé. Having lost their individual shapes and characteristics, they can no longer perform their biological functions.

The surface area of gills and the temperature of water: these are the two factors that, according to Daniel, explain fish growth and validate von Bertalanffy's equation. Not unlike Jared Diamond in his scientific works and books for the general public, Daniel offered an extremely detailed argument to defend his thesis. He dissected von Bertalanffy's equation, carefully analyzed its implications and the misunderstandings it had created, then wrote on for a hundred or so pages, embellishing his words with ninety-six equations edited by Dirk Reimers of the department of physical oceanography.

The subject matter is intellectually demanding and the debate could have been terribly boring, but Daniel's style is clear, powerful, and sometimes even funny. He demonstrates an astonishing mastery over a vast theoretical and empirical body of literature with consequences that go well beyond the growth phenomenon. Daniel, who had written his master's thesis in German, managed—not without difficulty—to get permission to write his doctoral thesis in English. It was his first big project in the language, but his style is perfectly fluid. As if by osmosis, he had absorbed all the writing techniques he needed from his extensive reading in the language. In his thesis, he blended them into a style all his own, hammering out the text on his typewriter with all ten fingers. Proud of his results and unbeatable on the subject, Daniel defended his thesis* brilliantly in May of 1979, then published a condensed version in 1981. The scientific review he chose was a low-profile one and the lack of response proved deafening. Daniel, sincerely convinced that he had made a major conceptual breakthrough, was shocked to realize that his colleagues had absolutely no interest in his theoretical work: he hadn't sparked even the smallest controversy. Swallowing his disappointment, he refused to abandon what he already considered to be his most important scientific breakthrough.

THE YEAR 1977 was a decisive one for Daniel's career and his intellectual growth, but it was also marked by two major events in his private life: Elizabeth Kambey, pregnant when they separated in Indonesia, gave birth to a son named Ilya, far away in Jakarta—and Sandra Wade entered his life.

I had never heard of Sandra until the spring of 2015. In order to work with Daniel and his team, I was spending four months in Vancouver with my family. The city, and the University of British Columbia (UBC) campus, had become a millionaires' playground, and the trip, productive and pleasant as it was, weighed heavily on our finances. To make ends meet, we had switched houses with a Canadian family, lending them our little house in Montpellier while we took theirs in Vancouver. We found ourselves in a beautiful turquoise abode built a hundred years ago with the wood of a giant torn from the old-growth forest—supposedly, just one of those cedars was enough to build fifty

---

* Titled "Gill Size and Temperature as Governing Factors in Fish Growth: A Generalization of Bertalanffy's Growth Formula."

houses. We soaked up the environmental and human history of the country. In Western Canada, though, that history is a depressing one, emblematic of the destruction of natural resources and the genocide of the First Peoples—even Vancouver Island, so lauded by the guidebooks, has lost over 90 percent of its old-growth forest.

"Canadians are generally indifferent when it comes to the environment," Daniel told my wife, Bénédicte, when she interviewed him for the French press. During our visit, he was more overscheduled than a politician—as usual—but accepted our invitation to dinner, adding, "I'll bring Sandra Wade." "Typical," I thought to myself, "the alpha male has started a new life with some young woman. I'm curious to meet her." Daniel arrived with terrifying punctuality accompanied by a quiet, distinguished Black woman his own age. We were all a bit shy and Daniel dominated the conversation. He made quite an impression on my adopted daughter, Ambalika, and his mustache and big glasses reminded her of her Indian father. Daniel gave her a serious look and declared, "In your career, nothing will be as important as your ability to write." There's nothing like an authority figure to make a mark on the teenage mind, and I am eternally grateful to Daniel for sharing that piece of wisdom. Sandra, for her part, spoke now and again in a lovely Southern accent, but remained mostly silent. Clearly, it's best to speak with her when her talkative husband isn't around.

I manage to do just that about a year later. Sandra has come to Europe with her sister Cheryl and we eat lunch together in Brussels on a warm Sunday in May. Her perfume reminds me of someone, but I can't decide if it is one of my favorite aunts or the therapist who treated my dyslexia when I was eight. Sandra sits comfortably, knitting while she tells me her story with a slight drawl evocative of long, quiet evenings. But she also observes me with a piercing clear-blue gaze that brooks no argument.

Sandra was born two months before Daniel into a family from Little Rock, Arkansas. She is the eldest of three sisters. They moved from place to place according to the assignments of their career soldier father, including to Darmstadt in Germany, where the family lived on the American military base from 1956 to 1960. "We traveled a lot in Europe. Our parents exposed us to as much as they could, whereas a lot of the American [military] parents did not. Planning the trip was always a part of it; we used to go to the library to look at maps," Sandra recalls. They hit the road often and visited, among other things, Expo 58 in Brussels. "Of course, we were a curiosity—most Europeans hadn't seen Black people before, but it didn't bother us." "That's true," Daniel adds when I talk to him about his in-laws. "They had a lot more freedom of movement in Europe than in the US. For a Black family, the United States was hostile territory—you had to plan your trip as if you were crossing the Sahara, identifying the oases in advance so that you could stop safely."

In Darmstadt, the Wade sisters went to the US Army school and learned piano and German. Back in the United States, their parents settled in California, never returning to Arkansas. They were some of the five million black Americans who, starting in the 1940s, left the impoverished, racist South to look for "the warmth of other suns."* Mrs. Wade was a stickler for her daughters' education and their independence—her watchword was "Always have your fare home." They all went to college and pursued postgraduate studies, and a new Black intelligentsia was born. For Sandra, that would mean a bachelor's degree in sociology from Berkeley, then a master's degree in political science from the Monterey Institute for International Studies. One of her professors insisted that she pursue her doctorate at the London School of Economics under the direction of Ralph Miliband and wrote to his British colleagues to ensure that everything would go according to plan.

Sandra moved to London in the autumn of 1970 and stayed there for two years. Her sister Cheryl joined her in 1971, then returned to Seattle to complete a PhD thesis on marmoset behavior at the University of Washington. Since his visit to California in the summer of 1969,

---

* Isabel Wilkerson, *The Warmth of Other Suns: The Epic Story of America's Great Migration* (New York: Vintage, 2010). Winner of the Pulitzer Prize.

Daniel had continued to correspond with the Wade family from time to time, and Mrs. Wade kept her daughters up to date on the wanderings of this odd bird. In the summer of 1971, he came to England for a visit. When he arrived at Piccadilly Circus, he tried to call Sandra but couldn't get the phone in one of the famous red telephone boxes to work for him. He walked outside to ask for help—leaving his wallet inside just long enough for it to disappear, along with all of his money. Mortified, Daniel borrowed some cash from Sandra, and his hundred-hour contract with Petz Arntz later helped clear the debt.

After London, Sandra got a job as an associate professor of political science at San Francisco State University, where she taught from 1973 to 1977 while working on a doctoral thesis at the London School of Economics on the philosophy of Malcolm X. Her research brought her back to London in the summer of 1977, then to Hamburg to visit the family of a German colleague she had met in California. There are only about sixty miles between Kiel and Hamburg, and it seemed natural that Daniel should come down for a courtesy visit. Everyone waited up for him, but he arrived hours late—he had simply lost the address. Sometime later, Sandra went to Kiel, "and the rest is history," as the discreet British saying goes.

Sandra, for her part, tells me the following: "We decided that we were both 'young, gifted and Black' and we should get married. He asked me if I would, and I said, 'Yes, surely why not.' That's how that happened. It was just one of those very pragmatic things." She moved in with him in Kiel and never returned to San Francisco to finish her doctorate. Daniel had warned her that marrying him would mean eventually becoming little Ilya's mother, since he planned to reunite with his son, to which she replied, "That's nice." In their wedding photos, from December 8, 1978, Sandra is wearing a chic black suit and a pearl necklace with matching earrings. Her hair is short, shorter than Daniel's—he is bushy-haired and long-bearded, sporting a beige jacket over a dark sweater and white shirt. Sandra looks delighted and delightful behind her thick-framed glasses. Daniel doesn't smile—he never does in photos—but instead looks tenderly at his young wife.

"Winston was held up by his family in California, but he was proud of Daniel. For him, marrying one of the Wade sisters was the heist of the century," Sandra comments.

The young couple lived at number 15 Adalbertstraße, near Kiel's military port and not far from one of the city's most beautiful parks, which is a great place to sit outside during the summer months. They both remember it as a happy time. Even the cold, dark, rainy winters didn't get to Sandra, who started German classes at the university and refused to be cowed by ill-tempered shopkeepers. With a little money loaned to them by Petz Arntz, the Paulys bought a small black-and-white television, a portable model in a shiny red plastic case. The TV was their window on German society, and they studied it from their living room, watching popular shows such as the legendary drama *Crime Scene*.* The four episodes of the American miniseries *Holocaust* were shown on German television that year. "For the first time, they showed the life of a Jewish family during the war, with Nazis as supporting characters—it really shook people up," Daniel remembers.

The atmosphere was studious, and Daniel often stayed home to work on his fish growth database and write his thesis. Sandra read his manuscript but didn't make many corrections. "He just writes particularly well in English in scientific papers. I was surprised by that, but I think it's because he read so many." She did, however, come to her husband's rescue when he lost his temper with the oceanographic institute's computer: thousands of fish-related data points had to be typed out individually by hand in one interminable list. If there was a single mistake, he had to start over from scratch. After a series of failures, Daniel was ready to throw in the towel—it was Sandra who managed to do it without any mistakes, on only her second try.

---

* The German title is *Tatort*.

BIRTH OF A CAREER
IN THE PHILIPPINES

I T WAS THE period historians call the Second Cold War. After a few years of détente, relations between East and West broke down toward the end of the 1970s—all over the world, capitalists and communists were once again at each other's throats. But as the saying goes, "You don't start a revolution on a full stomach"*—or, to put it plainly, you're less likely to become a communist if you aren't dying of hunger. The United States understood this and went to work on food aid and development in poor countries with the help of the United Nations and the FAO, though they also used private funds. In 1977, part of the Rockefeller family's huge fortune went toward the creation of a new marine resource management organization in the Indo-Pacific region: ICLARM.†

Initially based in Hawaii, the organization quickly moved to Manila thanks to some very effective lobbying by President Marcos. The research center was well funded and charged its first director, Jack Marr, with recruiting a "tropicalized" dream team. Marr was already an experienced researcher who had written a seminal work on Californian sardine fisheries that documented their collapse in the 1950s. He had also been director of the Indo-Pacific Fisheries Council and

83

---

* Jules Vallès.
† International Center for Living Aquatic Resources Management, called the WorldFish Center since 2000.

of another large FAO program in the Indian Ocean. Because ICLARM needed the best and had money to spend, Marr invited a small number of young scientists to come stay in Manila for a few months—including Daniel, whom he had heard of during a trip to Indonesia.

The young doctoral student packed his bags and put his romance with Miss Wade on hold for a while to go to Asia, from mid-June to mid-August of 1978. "By inviting a young Black scientist, Jack Marr showed he had an attitude that was uncommon for a white American of his generation," Daniel notes, remembering that Marr welcomed him with a very simple to-do list: "You are going to develop a theory of tropical fisheries." Daniel was petrified. "I was scared so stiff, I seriously considered turning around and going back to Kiel." But instead he jumped in headfirst, reading up on tropical marine environments and consulting a few general works on ecology, including a textbook[1] by Robert Ricklefs, then a young professor at the University of Pennsylvania.

After digesting the material, Daniel wrote a summary of thirty-five pages, with seventy-five bibliographic references.[2] It is a tour de force, considering how quickly he turned it out, but it is important to note that today, such a summary would require reading several thousand publications, reports, and books. Daniel benefited from the gaps in his new field and the unburdened freedom of pioneers. His report included a review of the marine fisheries of Southeast Asia followed by a presentation of the mathematical methods available to ensure their management. After that, he included a case study on fisheries in the Gulf of Thailand and finally, a general road map for the management of tropical fisheries.

Jack Marr's evaluation was nothing if not thorough: he submitted the whole thing to several notable colleagues, including the formidable British scientist David Cushing. Daniel's text was approved without any major modifications and published directly by ICLARM in 1979. Unfortunately, as with his thesis, it was shared with only a very small network of researchers—and yet his paper quickly became a classic in its genre, and remains one of his most cited articles today. The text is written in

English, the style simple and elegant, a Pauly trademark that proved his high school language teachers in La Chaux-de-Fonds wrong. The tone is frank, particularly when it comes to the governments of Southeast Asian countries, which he very directly accuses of developing industrial fisheries for export despite the fact that artisanal fishing actually employs the most people and feeds the local population. He also points out a "serious conflict" between short- and long-term objectives, and particularly the temptation to make a lot of money quickly in the short term. This problem has notably affected the Gulf of Thailand, which, Daniel showed, was already severely overfished. Daniel also threw a few barbs at the FAO for not proposing methods suitable for the management of tropical fisheries.

In many ways, this seminal report founded a new discipline and set down the major themes that would become Daniel's obsession over the next four decades. Poring over the pages, you can see the key terms: "overfishing," "collapse of the stock," "artisanal fishery," "multitude of tropical species," "fish growth," "removal of larger fish then smaller ones," "mesh size," "underwater natural parks or similar (non-) uses of the resource." The author concludes: "Almost all problems that can occur in a fishery do occur in a tropical multispecies fishery," and he suggests establishing a vast research program in this area of study. Whereas most researchers faced with a scientific problem would ask for more measurements, more data collection, Daniel simply suggests producing a synthesis of existing knowledge and asks for "a large computer capable of rapid operation and ... therefore quite costly."

More than satisfied with young Pauly's work, Jack Marr offered him a one-year research contract with the potential for renewal. Daniel accepted, leaving Northern Germany with his wife immediately after his thesis defense to settle in Manila in July of 1979. Ferdinand Marcos had been in power since 1965, martial law had been in place since 1972, the size of the police force and the army had quadrupled, and tens of thousands of political dissidents were behind bars, including the famous opposition leader Ninoy Aquino. Gun culture had also become deeply ingrained in the Philippines after the United States' violent takeover of the country following the end of Spanish colonialism in the first half of the twentieth century. In sum, not a very welcoming setting for expatriate scientists, most of whom lived as though in a war zone. This was the case for Ziad Shehadeh, the genteel Palestinian researcher who took over the direction of ICLARM after Jack Marr left suddenly following a disagreement over funding.

"Ziad lived in a U-tube," recalls Roger Pullin, one of his colleagues from that time. "He lived on the tenth floor of an apartment building, and he would go down to the basement, where his driver picked him up in a big limousine and drove him to the basement of the Metrobank building, where he took the elevator up to the seventeenth floor, and repeated the process. That's how he lived. His family was also very scared and they would soak all their vegetables in potassium permanganate [to disinfect them]." His Lebanese wife eventually convinced Dr. Shehadeh to move to Kuwait, where the Iraqi invasion would make him regret leaving behind the peculiar charm of the Philippines. His departure was all the more unfortunate because everyone at ICLARM liked Shehadeh—a good man, always willing to listen to his colleagues. The boss was, however, slightly perplexed by Daniel, whom he found a little too spirited. Shehadeh discreetly contacted several European and American colleagues, whose responses partially reassured him about his new recruit's character. "In the end, it was my wife who won him over," Daniel remarks. "Ziad liked Sandra a lot—he even made sure our company car had an automatic transmission because he knew she was American."

DANIEL'S NEW JOB paid well, and for the first time, at thirty-three years old, he attained a degree of financial stability. He still felt insecure about his future, though, and threw himself into his new work body and soul, only coming home to sleep. Sandra found herself isolated, stuck inside after curfew in the middle of Asia, far from her family, her studies, and her political activities. But prepared for this moment by a lifetime of nomadism, she landed a job as a teacher at an international school, where she initially taught psychology. After that, she deliberately worked her way up the ladder, ultimately becoming the head of the social science department, where she specialized in European history. Almost four decades later, Sandra says she has nothing but good memories of Manila and especially of the friends she made at work.

With their respective jobs in the nice part of town, Daniel and Sandra moved up in the world. They rented a large penthouse on the top floor of a very chic building across from a golf course and, after hesitating for a long time, hired a maid and a driver. Sandra, who loves plants, decorated their huge terrace with leaves in every shade of green, helped along by the tropical rains. From the point of view of their Filipino coworkers, the Paulys lived like royalty—they even had an open-air pool. "But I wonder if Daniel ever dipped a toe in it," Cornelia Nauen muses. Though researchers from the Global North who work in developing countries often enjoy a more luxurious lifestyle, moving south can also mean sacrificing their ambitions. Accepting a job on a tropical island with an organization like ICLARM normally means turning one's back on the big universities. This wasn't the case for Daniel Pauly, however. He worked "like crazy" all through the 1980s despite the fact that he found himself isolated from the European and North American circles that so totally dominated research at the time.

Daniel had another good reason to push himself. In 1980, he "retrieved" Ilya from Jakarta during an assignment in Indonesia. His son was not in good health, asthmatic and covered in scabs. Meeting another abandoned child, this time his own son, was incredibly painful for Daniel. "But it was love at first sight—I bought him a little toy car, and we played together and understood each other right away. I did my best to make up for what had happened." Indeed, according to his friends at the time in Jakarta, Daniel had left Indonesia in a less-than-honorable fashion, but he made up for it later. Ilya was a happy child, and in Manila as well as in his French and American families, everyone fell in love with "the most adorable little boy."

The next year, Sandra gave birth to a baby girl. The Paulys named her Angela as a tribute to the famous African American activist. The kids grew up next to the pool, around coral reefs, the prince and princess of a tropical kingdom that stretched from the family home all the way to ICLARM by way of the international school and weekends at the beach, with occasional intercontinental flights to California and France, via Hawaii. A photo from that time shows the Paulys, now a family of four, standing under some palm trees on the beach: a peaceful postcard from their tropical exile. Daniel wanted the best for his children and didn't joke around with their education. He sent Ilya to the French school: Daniel's son needed to master the language of Voltaire as quickly as possible, which wasn't easy in Manila, where he mostly spoke English with his Filipino neighbors and with Sandra. Ilya had just turned five and Daniel, with his usual extremism, decided to send him to live in France for a few months with Gérard and Jocelyne Pauly in La Rochelle. Sandra remembers, "I had just spent eighteen months becoming his mom and I cried every tear in my body, but Daniel had made up his mind."

Luckily, the La Rochelle Paulys were warm and welcoming, and little Ilya remembers those three months fondly: "I took the airplane by myself. I remember certain things very distinctly—the layout of Gérard and Jocelyne's apartment, going fishing for sea snails on the Île de Ré. After a couple of weeks, I was speaking French. When I got back, it was tough—I had forgotten all my English. It was my little sister who helped me remember." Family life picked up a regular rhythm. When the Paulys weren't working, they took care of their children; their friends were their coworkers, and Daniel soon threw himself into his first big Filipino research campaign.

SAN MIGUEL BAY AND THE SOCIAL DIMENSION OF FISHERIES

O N A WARM January day in 2017, I sit with José Ingles—aka Jingles—in the cafeteria at the University of the Philippines Diliman, a sizable campus shaded by giant acacias on the north side of Manila. Aside from the university, built by the Americans a century ago, the golf courses are the only green spaces for miles around. Our spot on the terrace seems pleasantly calm, though listening to the interview later, I realize that the noise of the city nearly drowns out our voices.

I have been navigating through Manila's hellish traffic for the previous week—a megalopolis of 20 million people, it has one of the most densely populated centers on earth. Since the Paulys arrived here in 1970, the country's population has risen from 50 to 100 million, and Manila now possesses thirty towers over fifty stories tall. The city is in the throes of a construction frenzy, the financial bubble not having popped yet in that part of Asia. The country's political situation is still delicate, however. To the south, a lawless zone is controlled by ex-Maoists-cum-mafiosos and Islamic extremists. On its northeast maritime border, this archipelagic country stands toe-to-toe with the Chinese giant and its strategic ambitions. To make matters worse, the freshly elected President Duterte has recently decided to go against the United States, calling Barack Obama vulgar names in front of the world's television cameras. Duterte has also given new powers to the police, recalling the darkest hours of Marcos's reign. In the months before my visit in 2017, thousands of drug dealers and users were murdered in the streets, and the police have just admitted to kidnapping and killing a Korean businessman in order to extort 100,000 dollars from his widow. For a naturalist like me who is only ever happy in the wilderness, this environment proves testing. Luckily, the Filipinos I meet are extremely friendly and relaxed, like Jingles, who agrees to share his story.

"My parents are in the fishing business; my family used to own a fleet of trawlers. Naturally, I was attracted to ocean science, and at twenty-three, I was a master's student at the University of the Philippines and working as a research assistant part-time on oceanographic research expeditions. I knew San Miguel Bay well because my parents had bought a trawler there; I knew it was extremely productive, but also that it was heavily fished, maybe with some conflict between the trawlers and the more artisanal fisheries. With my classmate Dennis Pamulaklakin, we had this idea of developing a research proposal that was biological, social, and economic. At the time nobody was talking about multidisciplinary analyses, and we had the idea over a beer. Dennis knew about ICLARM, so we took our chances to submit a proposal to them."

A few weeks later, the two friends received a call from Ian Smith, the chief economist at ICLARM. To their surprise, Smith was interested, and he offered to launch the project quickly, with only a few minor changes. "From there on out, we were employees on the project, which doubled our salaries," Jingles says with a big smile. "The funniest part was, we had written our budget in Filipino pesos, but ICLARM approved it in US dollars!* Dennis and I laughed so hard." The two students stopped laughing, though, when their supervisor from the university, Antonio Mines, sensing a golden opportunity, declared himself ICLARM's sole partner on the project. Disgusted, Dennis soon left research, but Jingles helped implement their ideas in San Miguel Bay before being recruited by ICLARM to work on other things.

---

* At the time, one US dollar was worth about seven Philippine pesos.

Daniel Pauly, for his part, was named the project's scientific co-manager upon his arrival in Manila. He would ultimately become its word machine as well, composing most of the reports, even those where Antonio Mines or other members of the team appeared as first authors. Daniel enjoyed working on the project, however, one of the few in his career that he hadn't launched himself. In fact, San Miguel Bay, located in the Bicol Region 125 miles southeast of Manila, is one of the poorest parts of the country, and artisanal fishing is vital for many villagers there. Daniel was keenly aware of the destruction caused by industrial trawlers and wanted to better understand their ecological impact on fish stocks, but also their effects on local economies and social structures.

"The great thing was that we had hired research assistants from the region's villages, fourteen people in all. We could really talk to each other because there was no idiotic hierarchy in the way." Daniel didn't know it at the time, but the whole region was covertly controlled by Maoists. Jingles confirms, "Everything went smoothly because, as university students, we were also in the struggle, so the rebels treated us as comrades. This despite the fact that the NPA* didn't like foreign scientists."

The team's first assignment was to build an 860-square-foot bamboo house that would serve as their accommodation and mission headquarters on land given to them by the town of Castillo. Michael Vakily, a young German newly arrived from Kiel, was also among the pioneers and would work through the whole following year to estimate the size of the fish stocks in the bay. Daniel left Manila to visit the team regularly, speeding along the dirt roads as though they were German autobahns. In particular, he helped take the first systematic measurements of the depth and salinity in the bay, an expedition that earned him another tropical sunburn, just like in the Sakumo Lagoon. Michael learned quickly that, when going on a mission with Daniel, by land or by sea, it was best to bring along a compass, since the boss, more interested in conversation than navigation, almost always got

---

* New People's Army, a branch of the Filipino communist party with strong Maoist tendencies.

lost. One of these excursions, which took Ian Smith and Daniel to the distant limits of the bay, almost turned tragic: the two scientists stayed a little too long at a local fisherman's house because he offered them a cup of instant coffee, "liquid gold for a family with so few means; they give you a pinch of it in a mug of hot water, as though it were a 1933 grand cru, and you have to honor the drink, like we were on a terrace in Rome with Sophia Loren." The return trip across the bay was dangerous: "We had a tiny little boat, it got dark all of a sudden, like it always does in the tropics, and we ended up out at night in the middle of a storm with huge waves filling up the boat." Ian and Daniel were saved from drowning by one of the trawlers they had criticized so vehemently: "We didn't want to say too much about who we were—it was a sticky situation."

During their two-year run, research assistants interviewed hundreds of fishermen and their families, using a single questionnaire that covered every aspect of fishing, from fish biology to governance mechanisms. "It was unique at the time," Daniel recalls. "We were able to get a complete social and ecological picture." What would Daniel and his colleagues learn from their research when the study was finished, in 1982? Mainly that the muddy and shallow San Miguel Bay (320 square miles) was being overexploited by the fisheries: 80 percent of its exploitable biomass had disappeared, and sharks and rays were nothing but a memory. The catches now consisted of smaller fish, squid, and shrimp. The approximately twenty thousand metric tons caught each year was three or four times higher than the official figure given by the Filipino government, which minimized the seriousness of the problem. What's more, nearly a hundred trawlers were exploiting the entire area despite the fact that they were only supposed to fish the deeper waters near the mouth of the bay. Their motors ran on diesel, subsidized by the state, whereas the smaller craft used in artisanal fishing ran on gas, which was not subsidized. The trawlers took more than a third of the catch, which benefited thirty-five families, while the remaining two-thirds fed two thousand families of artisanal fishers. Five bosses owned half the trawlers and collected a quarter of the total revenue. Daniel comments, "With their money, they take care of their families, plus one or two mistresses, and the rest is spent in bars and brothels. These guys walk around in jeans and T-shirts like everyone else, including the poorest of the poor. It's camouflage for their industrial operations, kind of like the discretion of the French bourgeoisie."

The communist NPA had infiltrated the research team, but this evidence of social inequality appealed to them, and they mostly left Daniel and his colleagues alone. The researchers' results led to the usual reports and publications, but they also distributed them in the form of brochures in Tagalog (the national language) and Bikol (the local language), beautifully illustrated with photos and drawings. These documents didn't just point out problems; they also proposed a number of management scenarios, an innovative move at the time. "It is critically important in this fishery, as elsewhere," the documents conclude, "that a management partnership is forged between fishermen and local and national officials with responsibilities in the fisheries sector."[1] Such a management partnership was put into place and even worked for a time, but eventually it vanished into the sands "because it's the Philippines," Daniel comments with a sigh. "It was doomed from the start; we were looking for solutions, but at the time we were like a drunkard looking for his keys under a streetlight." In particular, the law dictated that in order to prevent fishers from harvesting very small fish, the mesh size of a fishing net was not to be less than two centimeters (about three-quarters of an inch), "which is already very small, and besides, everyone cheated by using several nets layered on top of each other." The San Miguel project wouldn't have much local impact, nor would it attract attention from the international research community, the publications having only a very modest distribution. But for Daniel, it was still an essential step: "I learned everything in San Miguel, about the connections between fisheries and politics, equity, justice, about the challenges of accessing data and the challenge of highlighting it in an interdisciplinary setting."

In San Miguel, but especially in the Bolinao area that fellow marine biologist John McManus had shown him, three hundred miles north of Manila, Daniel made an important observation: "At the time, the sociologists thought that young men wanted to get out of fishing, whereas in fact, more and more children from these big families were fighting over shrinking marine resources. In the villages we saw, the children had big bellies, like little Ethiopians during a famine. The men drank and smoked and ate the family's money, leaving nothing for their wives, who looked old and tired by thirty, surrounded by an army of malnourished children. It was obvious in the villages but, at the time, none of the development programs were taking it into account. I didn't write about it either because I didn't have the right intellectual vocabulary yet, but it's the same process that supplies the brothels in Manila: the young women are much more mobile. They leave to prostitute themselves, usually in the capital or near the American bases, but they stay extremely loyal to their families and continue sending money, which subsidizes fishing. This kind of mobility is inconceivable for the men who are just fine in the village with their friends, their cigarettes, and their beers. That's why I have never understood the anthropologists' obsession with fishermen and their nice traditions. Besides, what I saw in the Philippines, and then in a lot of other countries, is that there are artisanal fishers who aren't traditional fishers. They are poor people who have only recently arrived from the inland areas, for whom fishing is a last resort. I discovered this some years later on a fishing boat in Peru: I saw a man who looked like a sailor, but he couldn't tie a single knot. It turned out he was from the Altiplano. Same thing in Vietnam and Thailand, where the sailors are often mountain men, from Cambodia, for example. The anthropologists hadn't noticed them because these people weren't 'interesting,' compared to the Lebu in Senegal or the Badjao of the Philippines, who have lived on their boats for centuries."

Daniel formulated these thoughts a few years later, in a series of articles published starting in 1988, where he defines the term "Malthusian overfishing."[2] To understand this concept, we have to go back to 1798. In England, the very severe Reverend Thomas Malthus anonymously published the first version of *An Essay on the Principle of*

Population.[3] This pamphlet highlighted the fact that the human population was growing exponentially, much faster than agricultural yields. Recent history has proven Thomas Malthus right: the world's population keeps growing, although it is evident that humanity is living beyond its means. The public image of this visionary scientist remains negative, however. The Nazis used his ideas for their own ends, as they did with Darwin, to construct their demented doctrine on the supremacy of an imaginary Aryan race. Even today, asking your friends and neighbors if it's really a good idea to have one, two, three, or four children is a good way to ensure you won't be invited to their next dinner party. It is surprising, therefore, that a leftist intellectual and product of the student revolution like Daniel Pauly would turn to Thomas Malthus. Daniel is, however, a rationalist, an intellectual trait he shares with the old reverend, even if it's unlikely that the man would have had Pauly over for a cup of tea in his time. "We think we all know who the Reverend Thomas Robert Malthus (1766–1834) was," Daniel wrote in 1990, "an obscure, long-refuted English cleric who was nasty enough to suggest that the poor of his time couldn't (and shouldn't) be helped because the 'geometric' growth of their population would always outstrip the productive capacity of the resources available to them... Malthus... ends up being vindicated: agricultural production cannot, in the long run, keep up with unchecked population growth."[4]

Daniel proceeds to adapt this idea for fisheries:

> There are too many small-scale fishermen in Southeast Asia. Moreover, their numbers are rapidly growing (with a doubling time of less than 20 years), both because of their own children and because of increasing landlessness among farmers, to whom fishing becomes the occupation of last resort... To put it bluntly: if I were a fisherman, with no alternative employment opportunities, and I had to feed my two children by blowing up a fishing ground, I certainly would do so... the solution is clearly alternative employment opportunities, in villages and towns and cities, for young and not so young, for literates and illiterates, and

stabilization of populations. Without these? Forget about coral reefs and mangroves, and forget about having options for coastal resources management.[5]

Daniel's strongly worded prose hit a nerve, and his "Malthusian overfishing" idea garnered its share of attention. Well couched in scientific terms, this argument is one Daniel would come back to throughout his long career, meting out increasingly strong criticism of maritime sociologists and anthropologists for failing to produce research that is useful for policy making. This series of provocations culminated in an article published in 2006 and presented during an international conference organized by the Centre for Maritime Research in Amsterdam.[6]

"I told them what they needed to hear," Daniel insists. More specifically, he said: "These are political problems, and they ought to be informed mainly by social scientists, not biologists. Yet fisheries biologists ... share with fisheries economists the dubious privilege of being responsible for most of today's ideas on fisheries management. Other social scientists, notably anthropologists, have far less input."

For Daniel, this is the fault of sociologists who "neglect key quantitative variables (this is especially true for the catch of small-scale fisheries), and fail to propose and test models of social behavior of sufficient generality to be useful for policy making." The same goes for anthropologists, who love to study the particularities of certain traditional societies. Each researcher ends up focusing on the "local adaptation of a village and emphasizing its uniqueness vis-à-vis other villages." But "there is a need for social-science generalization which is presently not met, for example to formulate people-oriented and sustainable government policies for an entire region or an entire country." Daniel uses the aftermath of the 2004 tsunami that devastated many coastal regions in Southeast Asia as an example. Sociologists and anthropologists approved the development programs that sought to reconstruct the fishing fleets, arguing that traditions should be maintained. For Daniel, though, this was the wrong approach; they should have seized the opportunity to offer alternative economic activities, ones not based on pillaging marine resources. "We should instead be teaching them to repair bikes, sewing machines and water pumps," he concluded.[7]

TROPICAL
STATISTICS

D URING THE EARLY 1980S, Daniel was exploring the social universe of San Miguel Bay fishers, but his main obsession was with something else entirely. Since his initiatory journey in Indonesia, he had been searching for new techniques for determining the growth rate of the numerous species of fish caught by tropical fisheries. The statistical methods developed by scientists from temperate climates wouldn't work because they required that researchers already know the age of the fish in each size class. Daniel dug through the literature, armed with a clipboard and a pencil that he sharpened with an electric pencil sharpener, his favorite gadget at the time. A second pencil was usually stuck behind his ear in case he lost the first one. At ICLARM, he already had a solid reputation as a mad genius, often walking around barefoot or in his socks among his white-collared colleagues, jubilantly ignoring administrative constraints, and stirring his coffee with his coworkers' pens.

During one of his bibliographic orgies, Daniel discovered that the Danes, before becoming world-class specialists in determining age from otolith cross-sections, had developed much simpler methods. In particular, toward the end of the nineteenth century, C. G. Johannes Petersen, a young marine biologist at the University of Copenhagen, suggested finding the growth rate of fish by examining the size classes of individuals of the same species captured over time.

To visualize this method, let's imagine a large group of school-children that includes every grade from kindergarten to the end of high school. In this slightly abnormal human population, children are born during one or two brief periods of the year, and groups of different sizes appear very clearly: we call these "cohorts." So, sixth graders from a spring cohort would be slightly larger, on average, than their classmates from the autumn cohort. Each year, the school nurses measure the children's height, but the medical service is very disorganized and writes down neither the date the measurements were taken nor the name of the student or their grade level. Using this anonymous data, though, we can still count the number of students in each size class. For a given year, there will be more students in size classes that correspond to cohorts than for size classes that fall between cohorts. If we were to represent this annual tally as a diagram,* we would see a series of small hills representing more numerous size classes. Then, we can align the diagrams from several years on a piece of paper, one per line, with the oldest at the top and the most recent at the bottom.

Looking over this page, we could see how the small hills move toward the right from line to line: the large cohort of sixth graders born in the spring would have grown a few inches when they were measured in seventh grade, then in eighth. Growth is especially visible among younger individuals, so we needn't continue our measuring program after they leave for college. In any case, when the size of a given cohort stops progressing, this gives us the size of the average adult for a group of students, and the time it took them to reach their adult size allows us to calculate the mean annual growth rate.

In the nineteenth century, without computers, statistical methods were primitive: C. G. Johannes Petersen, who was interested in the growth of fish, not schoolchildren, filled up entire pages with hills and valleys obtained by measuring thousands of anonymous fish (of the same species) captured over the years in the same area. After that, he

---

* Called a "frequency histogram."

drew diagonal lines with a ruler from the top left to the bottom right to connect the hills that supposedly corresponded to different cohorts. This is how he determined the maximal size and the growth rate using an approximative, visual method. Petersen's approach was much used, though with some variation, until the middle of the twentieth century, when it fell out of favor because of its lack of precision. Daniel, who had resurrected von Bertalanffy, did the same for Petersen by making two major adjustments.

First, he developed a new method for "smoothing" the hills and valleys so that the height of the hills wouldn't be impacted by the number of fish measured. "At the time, I mentally smoothed anything I saw—mountains on the horizon, women's breasts... smoothing haunted me."

Next, he developed a statistical approach that would allow him to calculate growth rate and maximal size by "trial and error." First, Daniel used von Bertalanffy's equation to determine the approximate growth parameters that were supposedly valid for the species of fish being studied. He then compared the supposed growth curve with the hill-and-valley diagram: the goal was to obtain a curve that passed through the maximum number of summits while avoiding the valleys. Little by little, he adjusted the parameters to obtain the curve closest to the ideal path. Daniel didn't mention this explicitly in his work from the early 1980s, but the technique he used was Bayesian. Bayesian statistics* are fascinating because they are based on a huge amount of research about a notion that seems extremely unscientific: *a priori* assumptions (or "priors"). If we create somewhat of a caricature, we might define a Bayesian as "one who, vaguely expecting a horse and catching a glimpse of a donkey, strongly believes he has seen a mule."[†] In the same way, Daniel had a priori knowledge of the maximal size and growth rate of the fish he was studying and used a likelihood principle to arrive at the solution that seemed closest to the truth. This process of

---

* Named for the British reverend and statistician Thomas Bayes (1701–61).
† Quote attributed to Karl Pearson.

statistical trial and error replaced and perfected the more approximate graphing methods by using formal criteria. In his analytical approach, Daniel was likely influenced by his times, as Bayesian statistics made a big comeback in the 1980s. These methods required a lot of computation, however.

"Luckily, computers replaced the visual line-drawing method; today we would call it artificial intelligence," Daniel remembers. As had become his habit, he quickly forged an alliance with an expert—this time, a programmer, a Filipino named Noel David, who coded a computational routine in BASIC. The two colleagues named the program ELEFAN (for "Electronic Length Frequency Analysis"), with a wink and a nod that would become the ICLARM gang's trademark during the 1980s. The first version of the program even featured a line of three stylized elephants walking across the opening screen. The routine initially ran on the first microcomputers that replaced the outdated calculators at the end of the 1970s. These "little" contraptions, like the TRS-80, looked like huge cathode ray tube television sets, with matching keyboards as thick as shoeboxes and maximum memory capacities of forty-eight kilobytes (a million times smaller than a basic computer today). At the controls was Jingles, who, driven from the San Miguel project by Antonio Mines, had become Daniel's very first intern. "I already had some knowledge of BASIC, but FORTRAN was the 'in' thing at the time," Jingles recalls. The two were thick as thieves. Inspired by Daniel, Jingles dove into the fisheries literature and began trawling for data. "All of it was in storage in the basement. I spent maybe three months digging up information in the Bureau of Fisheries, in the dust and dirt and sweat. I was actually rebuilding all these datasets. I loaded them species by species into ELEFAN, and it took hours before the first values came out... I was often working late at night."

AT THE SAME time, Daniel taught Jingles the basics of fisheries biology: "Daniel's big thing was independence; he really wanted to make sure that he gave me everything I needed to keep learning by myself." This desire to help others to learn also led him to teach at the University of the Philippines, on top of his already-hectic schedule at ICLARM. He lived halfway between ICLARM and the Diliman campus, regularly zipping back and forth in traffic that was much lighter than the ubiquitous nightmare that characterizes Manila today. Daniel taught a single class: "Fish Population Dynamics," but his Filipino students wouldn't forget him easily. Jingles remembers Professor Pauly as "an alien, like a creature from another world totally obsessed with science, but who explained things well, without being condescending to his students." "Sometimes he would make a cutting remark, for which he would apologize . . . I never saw him reject anybody who would ask for help," Jingles adds, "and that is something that I noticed in my three, four years with him. The door of his office was always open, which is amazing for someone who wrote as much as he did."

Over the years, ELEFAN would be completed by increasingly complex modules. It would allow researchers to estimate the size, and therefore the age, at which fish should be caught to ensure that they have had time to reproduce and maintain the population. This minimum could then help researchers to determine which mesh sizes can be used to fish sustainably. For Daniel, "it became possible to work on things that were totally impossible before in the tropical setting; it was outrageously simple." In fact, the program used data that was easy to collect—just fish length measurements. These can be collected from a research vessel, but also in fish markets. When Pauly and David launched ELEFAN in the early 1980s, biology laboratories were already overflowing with length measurements stored up over several decades, and that data went to feed an increasingly hungry ELEFAN.

The program was such an immediate success, especially in developing countries, that Daniel and his colleagues soon found themselves traveling nonstop to host training sessions and work groups on the subject. But the criticism they received was also severe, especially from Western researchers. "Those guys came after me because they said it was too simple. A lot of people were against it. And then, after ten years, it became a normal thing," Daniel recalls. Some early critics focused on the fact that fish growth rates vary seasonally, which can interfere with the analysis. In fact, even if seasonality is not especially noticeable in tropical waters, it does have a very real incidence on the growth of different cohorts of young fish. "So, we had to do all sorts of supplemental analyses and write a lot of articles, but gradually, we fixed the seasonal problem."

What's more, Daniel and his team quickly realized that their computational routine could be applied to almost all marine organisms, not just fish. In particular, ELEFAN would later be used a great deal on crustaceans, for which it is difficult to determine age classes, but also for a whole crowd of mollusks, turtles, marine mammals, even lizards. Jingles remembers making hundreds of copies of the program on floppy disks, the ancestor of the CD-ROM, and sending them to other researchers in developing countries. "Daniel's passion, his obsession, was motivating his colleagues to attack the problem of tropical fish growth," comments Jingles. John Gulland, the grand master of fish stock estimation at the time, put Daniel on his guard, however: "It's not people using ELEFAN [properly] that I'm afraid of, it's the misuse of it," he said. In fact, as Jingles notes, "People [who don't understand the program] use it like a cookbook, even today; it's what goes in, and what goes out."

In 1982, Jingles received a tempting offer from the GTZ that included the possibility of doing a master's degree in the United Kingdom. Daniel was hurt, but quickly riposted with an offer of his own, complete with a program in Kiel. "I accepted because I could bring my girlfriend to Germany, and I thought that the weather in England was far colder. I was wrong on that second point!" laughs Jingles. He survived three years in Kiel, writing and defending his master's thesis in German, the first of many foreign students whom Daniel would send to freeze on the coast of the Baltic Sea.

A brilliant career followed for Jingles, with jobs at the University of the Philippines, then the World Wildlife Fund, which he left in 2015. "Right now, I'm working on fish farming in the Philippines, just trying to show people that with aquaculture you can help enhance the environment rather than destroying it," he says. "Here, the level of poverty is about 32 percent, but in the fisheries sector it's about 42 percent, and the Philippine government isn't doing anything to help."

A PACIFIC HEROINE

F ACED WITH JINGLES'S imminent departure, Daniel already fore-
saw trouble with running ELEFAN and began frantically searching
for a new assistant who had the necessary skills in biology, but
mostly in computer science. With the help of his colleague John
Munro, he interviewed a few candidates. On April 1, 1982, a slim young
woman of twenty with short brown hair cropped shorter on the sides
showed up at ICLARM's welcome desk on the seventeenth floor of the
super-chic Metrobank Plaza building. Before going to meet these foreign scientists, she had put on a light-blue satin dress, purchased for her
best friend's wedding, and high heels.

She had never seen an African American in person. Daniel, who
was three heads taller than she, with big hair to boot, seemed intimi-
dating, silent and impassive behind his big glasses. John Munro talked
like a professor, but his face was a little kinder. The two men weren't
wearing ties, which she found disappointing. They asked her to program a linear regression in BASIC. It was easy for her but, writing on
the board, she talked out loud to reassure herself. Without letting her
write the last line of code, Daniel interrupted to ask if she could start
right away, if possible that afternoon. "What a great April Fool's joke!"
she tells me, laughing. Daniel had just recruited his most loyal partner, a guardian angel who, thirty years later, still watches over him
discreetly. Maria Lourdes Palomares, whom everyone calls Deng, is
of Filipino, Spanish, and Chinese heritage. Her story is emblematic
of the destiny of many of Daniel Pauly's young colleagues from the
tropics, unfairly caught up in the turmoil of postcolonial history, far

from the comparatively insulated atmosphere of research laboratories in the Global North.

"In Manila, my grandparents' house was in the middle of the rice paddies; the closest neighbors were five hundred meters [about 550 yards] away. You could smell the sea, the farms; I still remember the smell of wet leaves, mixed with the scent of Christmas, churros dipped in chocolate." Deng grew up among frogs and insects in the family garden, in the company of Jacques the parakeet, who was green, yellow, and blue with a very strong beak, and her dog, Toutou. On Sundays, the whole family would take a horse and buggy, a mode of transportation that was still common in the Philippines in the late 1960s, and go down to the beach. Deng learned to swim in Manila Bay. "Today, I wouldn't even dip my toes in it," she laments. Her father, the son of a wealthy family ruined by the Second World War, started out as a mechanic, then became the personal assistant of a manager at the Rockefeller Foundation. Her mother owned three Jeepneys, the Filipino version of the bush taxi that would gradually leave the more romantic Spanish carriages behind in a cloud of exhaust fumes.

Deng was a daddy's girl—her father helped her with her homework and taught her drawing and calligraphy. Her mother, who had a head for figures, took care of her daughter's economic education, even sending her to exchange dollars sent by relatives in the United States. "I was able to follow the gradual fall of the Philippine peso during my whole childhood," Deng remembers. This went hand in hand with the political crisis to which the country was succumbing. "It started in 1969 after Marcos's reelection; there were protests everywhere with deaths. I was protected from all that because I was still a child, but I remember my mother crying and my father talking about troubled times." In school, Deng tells me soberly that she was left-handed and "precocious." When she was nine, her math teacher, Madame Ramos, asked her to take some extra tests. "Two days of exams—I was so bored!" Deng sighs. She skipped a few grades and landed at an elite school with a hundred other "precocious" students. But Deng missed her neighborhood school and her friends. To pass the time, she became the editor of the school's newspaper. "I was number one in the advanced class, and they'd promised me a medal. My parents were so proud, but there was never any ceremony. They told me it was because of martial law—I was furious."

In high school, which she started at eleven years old, Deng met two people who influenced her greatly. First, there was Ms. Buendia, the science teacher, who had completed a doctorate in the United States. "The first thing she taught us was the reproductive system—she said, 'Girls, you are the queens of your own bodies. Pregnancy isn't mysterious, it's just biology.'" Deng repeated all this to her family—much to the dismay of her very Catholic aunts: "She speaks with the devil's tongue; someone needs to pray for her, exorcise her!" they would say. In philosophy, Ms. Estrada asked her students to call her Roxy. She had been a journalist in a collective that published radical reports on Marcos in 1969. "Despite the martial law, she spoke very freely. She trusted us, but looking back, I realize she took incredible risks," says Deng. The class read Plato and other classics, but also studied a series titled "Political Thought" and Mao's Little Red Book.

In her free time, Deng devoured the works of Isaac Asimov, her favorite being *Foundation*. She usually went to school on foot, picking up a friend on the way. One day, the friend didn't show up, and Deng noticed a crowd in front of the neighborhood church. A priest was stirring up the mob, and Deng spotted her friend and the girl's family. The man of God called for the destruction of the "Reds" and, to emphasize his point, pulled a revolver out of his cassock and held it up like a relic. "That day, I lost my faith. In class, I talked to Roxy about it, and she told me, 'Don't worry, you've seen the light.'" Deng's parents loved their daughter, but they worried about her: "Your head isn't on straight; all those tests have ruined you!" they said. They weren't sure of what to do about her, either, so they let it go.

On her final exams, Deng finished third in all the Philippines, and she went to university at fifteen. She began in engineering, focusing on math, then transferred to the school of medicine after two years in the biology department. "Asimov inspired me to pick medicine, thinking all these new technologies could save lives." Studying medicine was expensive though, even with a scholarship. A rich aunt supported Deng at first, but then she passed away. Deng left the Diliman campus at the age of nineteen after several studious and politically active years. "As soon as I got to college, I started going to protests. My political activities didn't affect my grades, though, because I knew I needed to have a place in society if I wanted to change it."

About once a month, Deng and some of her classmates would go to a training camp in the mountains, usually for a long weekend. "The professors were the ones who circulated the information, and you had to collect a series of passwords and several contacts to know the place, the date, and the time." The camps were mostly centers of political discussion: "We wanted to understand how the world worked—we knew Marcos oppressed people, especially the poor, and we wanted to know what we could do to change that. Marcos and his goons still had the old Spanish feudal mentality that we wanted to get rid of. The camps were hosted by the NPA, and of course, I considered taking up arms, but Ms. Estrada's lessons on pacifism won out. It was the right choice; using violence doesn't work in the long term. You have to convince everyone, and that's a question of education."

Deng stuck to nonviolent protests, but she still ended up on the blacklist—enemies of the Marcos regime who feared for their lives. In 2015, she showed me her photo album from the time, with group photos of her among her classmates. "This one disappeared, this one was arrested and we never saw him again... him, too... her, too... and him." Deng probably owes her life to a police officer uncle. "I was on the front lines of a protest, and, all of a sudden, I found myself face-to-face with my uncle, armed with a nightstick like all his coworkers. They were ordered to charge, and my uncle stayed where he was, shocked. I managed to escape, but I felt sorry for my uncle; I didn't want to do that to him." The uncle in question warned her mother: "It's dangerous, what your daughter is doing—she could die." Panic-stricken, Mrs. Palomares went to see her daughter on the Diliman campus. Deng tried to reassure her: "Don't worry about it, nothing will happen." In the end, her uncle managed to make his niece's name disappear from the blacklist—no one knows how—and to protect her during the whole revolution. He died sometime later, under suspicious circumstances during a police raid. Deng can't help but think that he somehow sacrificed himself for her; his death, and the disappearance of her friends, still haunts her.

During her last year at the university, Deng felt hollow, anemic, like the floor had dropped from under her—she knew she wouldn't be able to finish her studies in medicine because there was no money. Her family worried about her: "What are we going to do with you, poor girl? We could try to get you a husband, you know, you're actually very pretty!" One of Deng's godmothers had just opened a computer science school, the first in the Philippines. "Back then, the computers still looked like big cabinets with magnetic bands. I was interested in all that stuff, and I thought maybe I could program machines to help with medical diagnoses." Deng learned the basics of computer science and got a job at a telecom company with help from her father, who worked there. "I lasted a week; it was deadly boring—I typed lines of code all day." In town, Deng ran into John McManus, an American in the Peace Corps and a coral reef specialist who had been her diving instructor during a summer class at the University of the Philippines. It was McManus who told her about the job at ICLARM, and she leapt at the chance.

WHEN DENG FIRST started at ICLARM, she met Jingles, whose shaggy rock-and-roll look she liked right away. But he was supposed to leave for Germany soon; within a week, he taught her the basics of ELEFAN and she became the program's new administrator. Daniel also showed up with a pile of books. "Here's what you need to know about fisheries biology—you have two weeks." Over the years, Deng helped perfect ELEFAN's routines. In particular, she was in charge of trying to estimate the sizes of fish populations exploited by fisheries. Daniel launched her onto the international stage as an ELEFAN expert and instructor. In fact, the FAO, mostly financed by the Danish government, began a million-dollar project in 1982 to create a network of experts in managing fish stocks in tropical countries. The FAO soon realized that ICLARM already had such a network, and their program relied on partnerships between instructors from Denmark and the Philippines throughout the 1980s, putting on fifteen two-week workshops.

"And that's how we ended up touring Asia, Africa, South America, and the Caribbean," Deng remembers. "I traveled with ten PCs and thirty calculators in big boxes. I arrived a week before the workshop to get everything ready, then stayed for a week afterward to put the data in order." The nearly one thousand participants were delighted: all they had to do was show up with raw data on the length of fish and other marine organisms caught in their respective countries. Daniel, Deng, and their colleagues helped them determine the growth parameters using ELEFAN, then calculate the rate of mortality from fishing to better understand how the stocks in question were being exploited. Along the way, everyone wrote a scientific article—"Before the data could get lost, which unfortunately happens a lot," Daniel notes. He loved these workshops, which allowed him to learn about the practices and policies of fisheries throughout tropical Asia, as well as in East Africa and the Caribbean, but also because he was a fan of quick, commando-style operations during which he could escape any familial or administrative constraints and work for twenty hours a day completing the analyses of each of the participants, often entirely rewriting their articles at night so that they could all be published together in a hefty report at the end of the training session.

Siebren Venema, who led the FAO program, noted in the introduction to one such tome, "After many years of despair... a new era began."[1] Deng was a little more wary. "At first it wasn't easy because Daniel and I were the only ones using ELEFAN, which created some tension between us and the Danes, who had their own methods." Luckily, Venema's friendly nature helped smooth things over. He had known Daniel since their respective stays in Indonesia in 1975: "When Daniel got into discussions, he could be quite fanatic. But he's a very good lecturer and he really got the participants at the training sessions excited. He also got into it with Per Sparre and the other Danes who were capable, high-level researchers, but not as quick in a debate as Daniel." For Daniel, "Siebren wasn't trying to compete with me; he saw himself as a project coordinator. We never got into those games of 'who can piss the farthest' like with some of my other colleagues, especially from the FAO."

These colleagues accepted ELEFAN with difficulty, even if its routines were quickly integrated into their own manuals. Venema confirms, "Back in Rome, my FAO colleagues were against Pauly and his methods, but John Gulland agreed to publish it through the FAO, so it was distributed all over the world and it became a big success." But the conflict ran deep, and, in 1987, Jorge Csirke, John Caddy, and Serge Garcia cosigned a thinly veiled critique of Daniel Pauly's calculation routines, which they called "quick and dirty."[2] They admitted condescendingly that ELEFAN could be "very useful" in developing countries where the more sophisticated techniques developed in the Global North "will not be available... before a few decades." Their chapter lists a series of problems and questions without offering any solutions, leaving it to Daniel to get his hands dirty.

The Pauly and Palomares Flying Circus ignored these jibes and continued its tour at a breakneck pace that sometimes drove Deng to the brink: "In 1988, we did a training course in Tanzania. It was my first assignment in Africa, and I went to Mwanza on Lake Victoria by myself in a little seven-seater tourism plane. At first, it was wonderful—the pilot flew around Kilimanjaro just for me—but once we were on the ground, I saw poverty like I'd never seen before. I mean, I'm from the Philippines, I know what a shantytown looks like, but I saw the state of the people standing in the street, and I visited the hospital—I didn't know it could be that bad. I didn't tell Daniel about this during the training course, but every night, I cried in my hotel room, and when I went home, I had trouble eating; I felt so pathetic."

A MAN OF LETTERS

T HE SEA IS calm and Jay Maclean patiently explains that we are
going to be getting into the water from the beach via a narrow
channel he maintains between reefs. Next, about five feet beneath
the surface, we'll take a reading on the antique thermometer that he
uses to measure the water temperature each day. After that, we'll be
free to enjoy the rest of our dive on the coral reef, though Jay—who
looks a bit like Woody Allen—thinks he will probably be too cold after
about forty-five minutes. At eighty degrees Fahrenheit, the water feels
deliciously warm to me, but I'm more accustomed to the fjords of the
North Atlantic.

The underwater show is splendid: we are in the Philippines' Verde
Island Passage, in the South China Sea. For coral reef experts, this
region is the epicenter of a golden triangle of biodiversity that stretches
from the northern Philippines to Bali and the Solomon Islands. A team
of Australian researchers has observed the area around Verde Island
for the last thirty years, counting—among a thousand other colorful
creatures—more than a hundred species of fish along the coastline that
Jay and I are exploring at a leisurely pace on this tropical January day in
2017. Looking for nudibranchs, moray eels, and sea turtles, I hold my
breath, free diving for much longer than is comfortable, telling myself
that I might never see anything like this again, surrounded by an underwater garden whose biodiversity is unmatched anywhere on the planet,
and in the company of one of my literary heroes.

Jay is a quiet man who still manages to talk about his feelings,
which makes him all the more charming. I know that the idea of being

interviewed makes him uncomfortable, but he has been generous enough to invite me to stay for a whole weekend. The wood and bamboo house that he shares with his wife, Margie, is three stories high, surrounded by tropical vegetation, with a view over the South China Sea. I sleep out on the veranda without a mosquito net, the nighttime breeze and a large fan enough to keep the little critters away from me. In the morning, the melancholy song of the bulbul and the smell of bacon wake me from a deep sleep, the first good night's rest I've had in weeks.

After our aquatic adventure and his third coffee, Jay lets his gaze drift over the horizon, avoiding eye contact, and tells me about his life. A native of southern Australia, he studied biology and initially worked for the fisheries management agency in Canberra, a snooze of a job that he soon left for New Guinea and a project developing pearl oysters. The work was fascinating, the country dangerous, and Jay enjoyed the adventure, but his wife—pregnant with their third child—did not. They returned to Sydney, where Jay joined the Commonwealth Scientific and Industrial Research Organisation (CSIRO). Outside of work, he played in a rock band, painted, and wrote short stories "just for himself" in a style tinted with his self-deprecating sense of humor and a very British sort of melancholy. Spotting his talent, some shrewd manager named him editor of the fisheries department's publications. "The previous person there acted as a post office; just took all the papers from the scientists and sent them off to the journals. When they came to me with their papers I said, 'I'm not gonna send this, it's not in the right format, it's just not good enough.' I liked working on the manuscripts and making them more readable." This kind of job has disappeared from most research institutions, but for many years, scientific writers were employed full-time, the idea being that science types can barely string three words together, which is sometimes true. In the United States, that was how Rachel Carson, author of *The Sea Around Us* and *Silent Spring*, initially earned a living while she wrote her now-famous books in her spare time.

Jay is a worthy follower of Carson's tradition: "The word got out that someone was helping with writing, and before long I had a room full of papers," he says. Daniel Pauly, during one of his whirlwind tours, passed through Sydney in 1979, soon after he started at ICLARM. As was his custom, he dropped by to nose around the local fisheries department and noticed a bearded man next to a chalkboard that contained a waiting list of about a dozen publications. "We could use someone like you in Manila!" Daniel exclaimed. Jay didn't hear anything else from Daniel, but a few months later, entirely by chance, he saw a job posting in the ICLARM bulletin that matched his talents suspiciously well. He applied and went to Manila for an interview. "I didn't want to move to Manila after that," admits Jay. "When you come from a country like Australia where everything is quiet, it was nerve-racking to see policemen with guns outside every building, quite the same as it is now. I was staying in the Mandarin Hotel, near the office, which was *the* big hotel in those days, and there was a shoot-out in the lobby. It was like the Wild West." Jay returned to Sydney and after not hearing anything for another two months, he received a simple message: "Come as soon as possible."

"I wasn't very motivated," Jay remembers, "but when I went back to Australia, the same week, there was a big shooting in the Supreme Court in Melbourne. That convinced me that no country was really safe." In the interim, Jay had separated from his wife, and, with no attachments, he left for the Philippines. At ICLARM, Jay and Daniel hit it off, to Jay's relief: "Daniel does not like small talk. He's very quick in judging people, sometimes wrong, usually right, and once he has established what you are, he'll either talk or walk away. Once he decides someone is a lightweight, he will move on. He's sociable when the topic interests him. But otherwise, he goes away and reads a book."

At ICLARM, Jay joined a lively team where Roger Pullin, the new director of aquaculture development programs, had also just been recruited. "The 1980s were magic, although the music wasn't as good as in the sixties," muses Roger, who grew up in the north of England in the middle of Beatlemania and founded ICLARM's rock band with Jay. "At ICLARM, we had amazing support staff—secretaries, program assistants, research assistants—all people who were not only very pleasant to work with, but were competent and hardworking."

Roger, who had left a tenured position at the University of Liverpool, never regretted his jump into the unknown. "We had far more support and freedom than in academia. It was very fertile, especially for Daniel," notes Roger. "Daniel was breaking new ground," Jay adds, "and finding new methods that people could apply to any fish in any place in the world." "Of course, we were all run ragged, working very hard, and applying for funding was a big part of the job," Roger admits. Their search proved fruitful, however: ICLARM soon received an annual budget of six million dollars, including two million for Roger's aquaculture programs in developing countries. Even then, there were already serious doubts about the capacity of wild fisheries to feed the ever-growing human population, and aquaculture was put forward as the miracle solution.

Jay, for his part, left the fish alone and got down to writing. "First I upgraded the ICLARM newsletter, making it more interesting by putting in some articles and by getting people to write different things." Jay also organized and edited three publication series: the ICLARM Technical Reports, ICLARM Studies and Reviews, and ICLARM Conference Proceedings. These thousands of pages would become the privileged means of distributing scientific texts produced through their research programs, including the one on San Miguel Bay. "Daniel was lucky that we had this publishing machine and a pretty good distribution system—plus, we were sending out all the stuff for free. We had a big audience, in a way, and that really helped Daniel a lot," Jay concludes. Meanwhile, Daniel, who had only published three texts per year in the 1970s, churned out ten in 1980, twenty-one in 1982, twenty-five in 1984, and thirty in 1987, most of them as first author.

Working alongside Daniel and Jay was another person of letters: Leticia Dizon, who had studied English literature at the University of the Philippines and joined ICLARM in 1977. She worked as an editorial assistant and program manager for Daniel. "Being Black in a US organization, you have something to prove. He just worked, worked, worked. But he was kind, and we adored him," she recalls. This opinion is backed up by everyone I meet during my research in the Philippines. According to Annabelle Cruz-Trinidad, his assistant starting in 1987, "Daniel used to give out really difficult tasks around five PM; the staff would dread that moment, but he would call us 'marvelous' in the most charming way. Even if he was pushing people too hard, they wouldn't mind because he was pushing himself, too, working harder than everyone." Jay adds, "When he wants something, he sometimes resorts to the injured-bird tactic, as he calls it himself, when his voice breaks to just that extent that it sets off your emotions, like a tuning fork on a sounding board."

During break time, philosophical debates ensued. "Daniel saw a book about angels on my desk," Annabelle recalls, "and after that, he always tried to get into discussions about God and the Pope, and how much he hates the Pope, but I respected him as an atheist, and as a person I knew he was wonderful. And I knew he was a leftist, but at the time he was a 'pure scientist.' It wasn't until later that he became more involved and started working with people, NGOs, who can communicate the message." They also talk to me about Daniel the jokester, though his colleagues didn't let themselves be outdone: Leticia was pregnant with her second child when Daniel, a young parent himself, gently poked her swollen belly. "'Don't do that, or I'll name it after you!' And this is why my son is called Paul Daniel," she tells me, laughing.

This good humor also included the group of coworkers with whom Daniel, Sandra, and their children spent their weekends by the sea, one of the few leisure activities Daniel allowed himself during their years in the Philippines. In the 1980s, their circle mainly included Jay and Roger, who had married island beauties Margie and Tessie— the latter being none other than the actress Tessie Tomas, who played Imelda Marcos onscreen and hosted a popular talk show on national television. Margie and Tessie's high-society flair brought out Daniel's dubious sense of humor: when Jay introduced Margie to him for the first time, at a reception, Daniel looked over her daring black dress and asked her if she was in mourning. Margie would learn to tame the beast, however, and they all met up in Anilao—a few hours south of Manila, on the Verde Island Passage—on a regular basis. The three families shared the rent on a group of bungalows, their base camp for hundreds of diving excursions. A weekend in Anilao usually involved four dives, two on Saturday and two on Sunday, before heading back to Manila over the dirt roads. Daniel only went on the morning dives, preferring to spend his afternoons in the shade, sitting on a rock on the beach. He wrote or edited one or several manuscripts, pinned firmly to his signature yellow clipboard, pencil flying across the page and one hand nervously twisting a lock of hair.

Daniel's first dive dates from his time in Kiel, but it was his Filipina student Annadel Cabanban who took him on his first tropical outing. "He lent me some money to register for my first diving course, and as a thank-you, I took him on his first dive. He was so focused on the technique that he barely looked at the fish. Later, he told me that when we reached twenty meters [sixty-six feet], he looked up from the depths and thought, 'Daniel Pauly, are you scared?'"

He would, however, look down his nose at the joys of coral reef diving for many years. Once, he gloomily told Jay that "all I see down there is conflict, fish spending their time chasing each other and fighting—I'm swimming through the tragedies of the deep." It took him some time to accept diving's recreational value—and to pass his certifications. As Jay tells it: "It goes without saying that Daniel's diving followed the same absentmindedness as his terrestrial pursuits. His near-legendary exploits sometimes brought him into close proximity with 'Z'—the fisheries scientists' term for 'total mortality.' This could include becoming separated from his buddies during dives and causing them to search the sea bed fruitlessly, while he was bobbing at the surface, having either forgotten to open the air valve properly, used an empty tank, or used a full tank in record time, or filled his buoyancy jacket with air by mistake, or simply because he left the group unannounced. Or he'd bring his son's or daughter's equipment by mistake, and their masks would be too small and would leak, or in any case, they did not have prescription lenses like his own, and he would have to be led through the dive like a blind man."

"But we love him," Jay concludes, "And I think I can speak for many people who have helped repair his broken wings from time to time in saying we still love him."

FISH STORIES
IN PERU

OCTOBER 24, 1978: NASA launched the Nimbus 7 from Vandenberg Air Force Base in California. The last of the Nimbus satellites, the first of which went into space in 1964, this 1,834-pound baby carried a number of sensors designed to observe terrestrial surfaces. Among them was a radiometer that, for the first time, measured the light reflected off the oceans. The more phytoplankton there are in the surface waters, the more they tend toward a blue-green color; these eyes in space could therefore map the productivity of the world's oceans. To make them easier to read, the blue-green shades Nimbus 7 measured were converted into a rainbow of colors ranging from blue to red.

The results were fascinating, and a copy of that sparkling planisphere soon found its way onto the wall of my childhood bedroom. Over the years, I learned that certain parts of the ocean are more productive than others. In particular, bright-red patches can be seen all along the coast of western and southern Africa, and from California to Peru, where one would have expected to find a blue desert like in the center of the Mediterranean. These unusual areas are "upwelling zones" where cold, nutrient-rich water rises to the surface. Drawn from ocean depths of two hundred to three hundred meters (650–980 feet), these sea-mineral cocktails boost the productivity of the photosynthetic microalgae that feeds the zooplankton, which then fatten up the small pelagic fish and the creatures that eat them, including carnivorous fish, birds, and marine mammals. Upwelling zones only make up about 5 percent of the ocean's surface, but they produce 25 percent of the fish caught by people.

The upwelling phenomenon is a fascinating one. My oceanography manuals explained it as follows: Imagine an intercontinental missile launched from the equator in the direction of the North Pole. During its flight, the Earth continues to turn on its axis, toward the right. The actual trajectory of the missile relative to the ground will not, therefore, go straight toward the north, but slightly northeast. This deflection phenomenon is called the Coriolis force, and it affects the trajectory of all moving objects that are not firmly attached to the Earth, pushing them toward the right in the northern hemisphere and toward the left in the southern hemisphere. Masses of air and water are also subject to the Coriolis force, which affects oceanic currents. In certain areas where winds push currents along the continents, these giant masses of water are forced away from the coast, which causes more water to be sucked up from the depths toward the surface.

Everything is bigger and richer in these upwelling zones. Even on the foreshore, the shellfish are gigantic—the limpets grow as big as my thumb, the abalone as big as my hand. In southern Africa, these abundant marine resources have fed humans for the last 100,000 years, as mounds of semi-fossilized shells and fish, bird, and whale bones left by the Khoisan people attest. Some believe that our amazing cerebral capacities may even be the result of copious early seafood consumption. Upwelling zones are therefore the cradle of humanity and have long nourished Indigenous peoples. They were also quickly spotted by European navigators, who used them to resupply. Some explorers, including Alexander von Humboldt in Peru, reported that clouds of ocean birds fed on sardines and anchoveta in these zones, leaving behind enormous amounts of guano, which is an excellent fertilizer. This led to a guano gold rush in the nineteenth century and the eventual exhaustion of the resource, which was replaced by synthetic fertilizers in the twentieth century. After that, it was mainly the unfathomable fish stocks in upwelling zones that became the object of industrial exploitation. Things happened quickly: in only two decades, the Americans cleaned out the schools of small pelagic fish in the upwelling off the California coast. The fishery collapsed a few years after John Steinbeck published his famous novel *Cannery Row*, in 1945.

In the early 1950s, certain California fish meal plants were taken apart, shipped to Peru, and rebuilt there without the knowledge of the local authorities. "If you can import whole plants and run them, despite the fact that they stink to high heaven, in secret, it means there is something seriously wrong," notes Daniel. Up to then, Peru had planned its use of marine resources around guano production: the seabirds that produced the natural fertilizer—around twenty million cormorants, boobies, pelicans, and penguins—were entirely protected, as was their food source, the anchoveta. Currently, however, it is the anchoveta themselves that have become Peru's main export in the form of fish meal used to feed pigs and chickens, and in aquaculture all over the world. The use of fresh anchoveta as a food source for Peruvians, many of whom are malnourished, is not considered a serious possibility, despite the fact that it is technically and culinarily possible.

Once fish meal plants moved into the Peruvian industrial landscape, a fisheries institute was founded: IMARPE, or Instituto del Mar del Perú. The institute is every bit as gigantic as the resources it studies. Since its creation, it has employed thousands of people, and today has a staff of seven hundred, even if many of those jobs are temporary. IMARPE's three research vessels are at sea three hundred days per year. Alongside associated land-based research stations, those ships gather every piece of accessible data on the marine environment, the fish, and their predators—day after day, year after year. A very Soviet-looking photograph shows this army of researchers in 2014, on the institute's fiftieth birthday: the lab techs are in white coats, the rest in dark suits. Most of them are men, but two women stand among the nine directors in the front row. For Arnaud Bertrand of the French IRD, who spent six years working with the Peruvians in Callao: "It's hard to do any better: it's obviously the most studied marine ecosystem in the world."

The oceanographic mechanisms of coastal Peru are all the more fascinating because, from time to time, they stall. The Indigenous peoples of South America were familiar with the phenomenon, which their Christian descendants baptized "El Niño" after the baby Jesus. In fact, the whole thing starts around Christmas: the trade winds that cause the Humboldt Current and the Peruvian upwelling system weaken, and tropical waters from the northwest take their place. These warm, nutrient-poor waters close like a lid over the coastal areas of Peru, stopping the normally abundant plankton production. This very thing happened in 1972–73, and anchoveta catches, which had been exceeding sixteen million metric tons per year (25 percent of the world catch at the time) collapsed to two million metric tons per year. Some marine animals managed to flee south, but many died on the spot, including thousands of seabirds and marine mammals who were totally dependent on the anchoveta. Beyond the tropicalization of the surface waters off Peru and a drastic reduction in marine life, El Niño is also associated with heavy rains in Peru's normally arid coastal areas and many climatic irregularities (droughts, flooding, storms) in other parts of the world. This global phenomenon, also called ENSO, for El Niño–Southern Oscillation, has been a source of fascination for oceanographers for decades, especially since it seems to be increasing in frequency and intensity with the planet's temperature rising as a result of human activity.

Researchers in the field of evolutionary ecology dream of working in the Galápagos Islands, but for oceanographers, there is only one mecca: the Humboldt Current and the Peruvian upwelling system that is subject to El Niño. The crème de la crème of marine scientists tends to flock to Callao, and fisheries biologists are no exception. "Big names like John Gulland, Bill Ricker, John Walsh, Garth Murphy, and other Americans used Peruvian data and published their papers in prestigious reviews without including the local researchers," Daniel explains. It was Petz Arntz, for whom he had invented the mud-sorting machine years earlier, who opened IMARPE's doors for Daniel. Petz was running a development program financed by Germany, focusing on Chile and Peru. Since Kiel, he had kept an eye on Daniel's career. "Petz had scoffed at my thesis on oxygen," Daniel remembers, "but he recognized the utility of ELEFAN right away."

As a result, Daniel was invited to come for a few weeks at the end of 1981 to teach, but also to evaluate the potential of the available information on the growth of small pelagic Peruvian fish. "IMARPE was housed in a huge building, and out front, there was a collection of rusty old cars, vehicles donated by various development programs over the years; it was like a museum of automobile history from the 1950s, '60s, and '70s. The director of the institute changed every time there was a political crisis, and IMARPE was actually run by four very distinguished women, one of whom was Isabel Tsukayama." This Peruvian woman of Japanese heritage had been working on fisheries for the last twenty years. She ruled her team with an iron fist and her previous experiences with *gringos* had not always been positive ones. "I found her charming and I think the feeling was mutual," Daniel explains seriously. "Isabel gave me whole stacks of documents containing length frequency data, which I took back to Manila with me." At ICLARM, it was Deng Palomares's job to feed the data through ELEFAN in order to determine seasonal growth fluctuations and the biomass of Peruvian anchoveta from 1961 to 1979.

"The data was extraordinary, with very clear growth patterns," Daniel remembers, digging up the article on the subject and showing me the graphics proudly. One of the biggest advances made by that analysis was to put forward not one but twelve annual estimations of anchoveta biomass, which would make it possible to test the impact of environmental variability on the population. "I went back to Peru with a finished article and offered to make Isabel the first author. She turned it down, but she was touched." After months of transpacific correspondence between Callao and Manila, Deng finally met her Peruvian colleague: "It was strange to meet another Asian Spanish speaker on the other side of the world. Isabel was even shorter than I am, pretty, an elegant forty-year-old, very smartly dressed. There was a real intellectual compatibility between her and Daniel even though she was very conventional and he was a lot more rock and roll."

After this promising start, Daniel suggested to Isabel that they coedit a book dedicated to the ecology of anchoveta in the Peruvian upwelling system over three decades, starting in the 1950s. She accepted and convinced her colleagues at IMARPE to get on board and collaborate with Pauly's team at ICLARM. A long series of Peruvian expeditions followed, and Daniel set himself to learning Spanish: "I speak it fluently, but I make tons of mistakes. It's not pretty, but it's intelligible." The book, mostly financed by Petz Arntz's program, had two main goals: "IMARPE didn't have a procedure in place to make those huge mountains of data available," Daniel notes. He therefore offered to include all of it in the volume, which would include dozens and dozens of tables crammed full of figures with everything from wind speed to the volume of anchoveta caught, and even the number of nesting cormorants on the guano islands, month by month, over a period of thirty years.

In the pre-internet era, making this much scientific information available was very unusual. But it was part of Daniel's philosophy, as he explained in the introduction to his first Peruvian book and again a few years later in another short chapter.[1] For him, any analysis of ecological processes must include a historical dimension. It is therefore essential to compile data over as long a period as possible in order to avoid what he calls "reproducibility erosion." From his studies in Kiel, Daniel also knew that physical oceanographers had been exchanging scientific data freely for over a century, a practice that greatly benefited their discipline. He wagered everything on the application of this precept within fisheries biology. Scientifically, one of the objectives of this huge analysis centered on the Peruvian anchoveta was to test Reuben Lasker's hypothesis,[2] according to which storm activity determines the proportion of anchoveta larvae that survive and eventually reach adulthood. Lasker suggested that the storms cause a strong disruption in the surface waters, where the plankton that feed the fish grow.

A washing-machine-like effect then shakes up both predator and prey, which can only find each other during periods of relative calm.

When he arrived in Lima for one of their sessions, Daniel didn't set foot in a hotel. He preferred to stay "with family" at Claudia Wosnitza and Jaime Mendo's house. Claudia is one of Daniel's old classmates from Germany, who went to Peru for her thesis in the 1970s. Her husband, Jaime, also a marine biologist, comes from a village on the Peruvian coast where the landscape resembles the lunar surface. He had met Daniel in Kiel during his master's degree in 1979. "The weeks with Daniel were wonderful, but also tiring for Claudia and me," Jaime remembers. "He would get off the airplane with a new manuscript written in pencil on little pieces of paper, and the only non-scientific part of our conversations was 'hello.'" Jaime was sent on a data-mining expedition in his colleagues' offices at IMARPE to search for any and all historical information on anchoveta and their environment, with a secondary diplomatic mission of convincing the data's owners to participate in Pauly and Tsukayama's big synthesis. "I spent months digging up sheets of information, half-eaten by insects and mice, from the backs of cabinets, and inputting the data. But it was worth it—in the end, we got a complete, 360-degree view of the anchoveta's little world over three decades; it was unprecedented."

This first volume devoted to Peru kept Daniel busy until 1986. He supervised all the analyses, encouraged his Peruvian colleagues and coauthors, and edited most of their writing, besides writing the introduction and conclusion to the 351-page volume himself. The scientific message was clear: since the collapse of the anchoveta stocks in 1972–73, birds and marine animals were no longer the main consumers of anchoveta—they had been replaced by carnivorous fish and human fishers. "It is not the sea mammals which threaten the Peruvian pelagic fisheries, but rather the converse," Daniel concludes.[3]

He began preparing for the next phase by noting that the first volume did not take socioeconomic factors sufficiently into account, and that it should be considered as simply a first step "toward an integration of what is now known on the dynamics of the fishes off Peru into a large-scale simulation model that could be used to help formulate a comprehensive fishery management plan for that system."[4] This sketch would be the starting point for the second volume, coedited by Daniel Pauly, Peter Muck, Jaime Mendo, and Isabel Tsukayama.[5] This new 438-page ode to Peru's marine environment went much further in its ecological conclusions. In particular, Daniel and Deng wrote a chapter on fluctuations in the anchoveta population: "The recruitment collapse of 1971 appears to have occurred before the onset of the 1972–1973 El Niño event . . . [this] implies that El Niño . . . cannot have been the cause of the collapse of the fishery." In an excellent synthesis at the end of the volume, Peter Muck describes an ecosystem that has been profoundly transformed since the early 1970s, with warmer surface temperatures reducing the productivity of coastal waters and favoring species that compete with the anchoveta, whose populations had had a hard time recovering. According to Muck, it was mainly the fisheries that prevented the anchoveta stock from regenerating: "The Peruvian fishing industry has now at its disposal only half of the reserves it [had] 30 years ago, when the Peruvian population was half its present size." He argues for the need to better take into account natural fluctuations in anchoveta stocks and for the use of fresh fish as a food source by local populations, rather than as fodder for fish meal destined for export.

The two volumes became landmarks. According to Arnaud Bertrand of the IRD, they are still "the reference." "These books are still in all the IMARPE offices today," he confirms. Bertrand believes that Peru responded to the findings in the correct fashion: quotas were put in place that allowed the stocks to regenerate; indeed, the volume of the catch reached ten million metric tons annually in the 1990s, the same levels as in the 1960s. "Once the quota was set, the fishing fleet started being tracked from day to day. The scientific director of IMARPE is in direct contact with the vice minister of fisheries, and if there is the slightest problem, the fishery is closed the very next day. When it comes to industrial fishing, it's hard to think of a more adaptive management scheme. Of course, there are always problems with reporting, undeclared landings, and artisanal fishing isn't tracked."

Jaime Mendo, whose thesis Daniel directed and who has since become a professor at the National Agrarian University–La Molina in Peru, is more critical: "Even today, we're still trying to use environmental variability, and specifically El Niño or other large-scale climatic perturbations, to explain fluctuations in anchoveta stocks. The impact of fishing isn't taken sufficiently into account. Specifically, too many young anchoveta are being caught, then thrown back into the sea to make room for larger fish in the holds of the boats, so we lose billions of dollars and future breeders." In fact, whereas overfishing was clearly mentioned in the two volumes edited by Pauly and his colleagues in the 1980s, this terminology was absent in the special volume that Arnaud Bertrand and his colleagues devoted to the Peruvian upwelling system in 2008.[6] Only Pierre Fréon of the IRD brought up the fact that the anchoveta fishing fleet was too big compared to the state of the resource it exploits, and that, despite the regulations in place, the stocks were, practically speaking, accessible to anyone. He also reminded his readers that, for the period from 1999 to 2005, only 6 percent of the fishery's catch had been used to feed local populations (of which 40 percent live in poverty), and that the Peruvian government's neoliberal policies had facilitated the overcapacity of its fishing fleet, leading to a ticking time bomb for the fishing sector. Indeed, as Fréon wrote, "A reduction of the fishing and processing capacity and measures to decrease the investment lag are recommended, to limit the social, economic and political tensions that will result from the expected decrease in stock abundance."[7]

Patricia Majluf confirms this and goes further still. I decided to interview this cosmopolitan Peruvian because she has known Daniel since he began working in Lima: "In 1983, I had just started my thesis on sea lions in conjunction with Cambridge University, and I met Daniel in Petz Arntz's office at IMARPE. He literally popped out from behind a pile of books and asked, 'Who are you?' We started having lunch together, and I would tell him about behavioral ecology and he would tell me about his mathematical models for fish. He always had several projects going at once, collecting data from everywhere, so I called him *El Pulpo* [The Octopus]. At the time, Peru was going through a large El Niño event. In the 1990s and early 2000s, the stocks came back, and in 2006, the public officials very judiciously decided that they needed to always leave five million metric tons of free anchoveta in the system. It was an excellent decision from an ecological perspective, but also economically because the prices went up—everyone was happy. But what can you do? People are too greedy, and they'll do anything to circumvent the rules. On the one side, IMARPE pieces together data to make it look like the five million metric tons are still there, and on the other side, catches are underreported. The whole chain of command is corrupt; fraud has become part of the system."

Patricia, who currently runs the Peruvian branch of the NGO Oceana, forged an alliance with Daniel and a few others to promote eating anchoveta, "the most nutritious fish in the world." This was a big challenge because, as she puts it, "All the infrastructure was built around producing fish meal for export. But in 2006, we managed to save 130,000 metric tons for human consumption." The road is still long, though, and Peruvian legislation does not favor better management of the anchoveta stocks. "The word 'overfishing' isn't in the lawbooks," Patricia tells me. "The three official stages of a fishery are: underexploited, fully exploited, and reconstituted, meaning that a 'reconstituted' stock necessarily comes after a drop due to natural environmental cycles, not overfishing!"

The debate around the management of Peruvian fisheries remained intense, and it was a problem to which Daniel would return. Jaime Mendo recalls, "Daniel was invited to Lima for a conference in 2006. He was excited to see everybody, and, at Patricia Majluf's suggestion, we decided to jump in and produce a third book on the Peruvian upwelling system." They chose a structure similar to the one used in the 1980s, putting the emphasis on making IMARPE's data available, and Daniel arrived armed to the teeth with a new methodological arsenal, hoping to influence the powers that be. Their analyses began in 2009 and 2011, but times had changed since the 1980s. Isabel Tsukayama, a guardian angel who died young, was no longer there to protect Daniel, and he was saddled with a Peruvian editorial team with whom he soon had many disagreements. "It doesn't take much to upset Peruvian managers," noted Daniel, who did not get along at all well with the director of IMARPE, whom he nicknamed *Pistolero* after he saw the man pull a revolver on a stubborn underling. In 2011, things came to a head and the divorce was made final. "We already had two-thirds of a book with analyses for a sixty-year period—two years of work for nothing. IMARPE is the only place I've ever failed like that. After the success of the 1980s, it was my Waterloo," laments Daniel. And he is not the only one: at the IRD and in Peru, everyone seems to feel bad about the outcome. For Jaime Mendo, it was "clearly a political problem."

Back to the 1980s: while Daniel was immersed in his Peruvian project, between 1983 and 1984, ICLARM was experiencing major financial problems. "The Rockefeller Foundation had simply decided to stop financing programs to fight hunger in the third world," Daniel remembers, "and we had to find a new funder." Roger Pullin recalls that "despite Daniel's growing notoriety and his air of confidence, the idea of finding himself unemployed made him very nervous, probably because his own path to success had been so difficult." A job opened up at the FAO, and Daniel promptly applied. They turned him down, though, and the Rome-based institution hired a French tropical tuna specialist instead. Suddenly without secure funding, he returned to Kiel in 1984 with a contract pieced together by his allies at the oceanographic institute that gave him a few classes to teach and six months to obtain his "habilitation to direct research."

In Germany, as in France, the habilitation to direct research* is the highest-level diploma available, one step beyond a doctorate, conferring on the bearer the right to direct theses and apply for university professorships. To obtain it, one must produce a written document and pass an oral exam. Often, candidates simply present a collection of their publications, along with a few pages of introduction and a

---

* Called the HDR or *Habilitation à diriger des recherches* in French. (TN)

conclusion. But Daniel translated and rewrote everything into perfectly fluent German, churning out 171 pages on his typewriter and adding accents and other special characters by hand. He begins by reminding the reader that tropical fisheries represent 30 percent of the global catch, and often provide 50 percent of the protein available in poor countries. After that comes a vehement critique of the notion of development, which, according to Daniel, too often leads to the destruction of natural resources. Then, Daniel rolls out an elaborate argument developed during his long years in Asia on the inadequacy in tropical climates of methods created by Western researchers. This naturally leads to a very detailed and strongly argued presentation, supported by numerous examples, of ELEFAN's different modules, which make it possible to study growth and mortality parameters in fish and other tropical marine organisms. He concludes by asking what would be the best method for estimating the proportion of young fish that grow to adulthood,* and presents the latest conceptual developments in modeling, taking into account several species of fish simultaneously. With his characteristic dry humor, Daniel notes, "This document, in its entirety, only covers aspects of population dynamics that correspond with the author's domains of expertise. He can only hope that these are as significant as he believes them to be." In the same vein, further down: "This equation is not very elegant and looks complicated, but it only contains five variables." Daniel likely allowed himself these kinds of liberties because he knew his scientific dossier was already more than enough.

At the end of 1984, Daniel went into his oral examination with the same confidence and found himself face to face with a whole jury full of bigwig researchers. According to his colleagues present that day, Daniel shone once again by his impertinence. "One physicist didn't agree with some parts of the multispecies models I had recommended," Daniel recalls. "In front of all the professors of the college of sciences, I told him to his face that physics is easy compared to biology."

---

* What experts call "recruitment."

Daniel bet on the fact that most of the votes he needed were already won. "I could take the risk of putting this guy in his place, which is probably something no one had done before." Daniel's friends still held their breath for him, though, and they think it was probably Gotthilf Hempel's presence that saved him once again.

Habilitation in hand, Daniel left Germany, returning to his job at ICLARM, where the financial situation had stabilized, and continued to work like a madman. Alan Longhurst contacted him, offering to coauthor a book on tropical ocean ecology. Professor Longhurst was also the director of the Bedford Institute of Oceanography in Halifax, Canada. Twenty years older than Daniel, he was an international authority on plankton and was specifically interested in the distribution of phyto- and zooplankton on a planetary scale. This work had led him to define and classify the great marine regions according to their ecological characteristics,[8] zones Daniel suggested naming "Longhurst areas." This global approach made sense to Daniel: "You have to look for the big picture before the details—parochialism is dangerous for everyone, including fisheries biologists," he says. "I've always been struck by Alan's encyclopedic knowledge," he adds, "though he acts as though it isn't important or it's superficial, but, of course, it isn't."

In 1985, Longhurst was nearing retirement. "He was very disappointed," Daniel remembers, "because the Canadian oceanographic research flotilla was being dismantled—it's the classic scenario when bureaucrats take control and destroy everything." Alan sought refuge in writing and soon moved to Cajarc, in southwestern France, where he opened an art gallery with his wife. Daniel accepted the writing partnership despite his numerous more modest occupations. He mainly agreed to take on the project out of admiration for Alan, but also because he knew that it was the best way to firmly anchor his concepts and methods in the scientific literature. Alan promised him "that it would be a brief synthesis." Daniel flew back and forth to Halifax two or three times. Ultimately, the duo produced 407 pages with 675 bibliographic references, including one from 1927 relative to the work of Théodore Monod. Their book appeared in 1987[9] with

Alan as the architect, writing eight of the ten chapters. The authors' stated objective was to "break the myth of tropical oceans dominated by coral reefs and mangroves." In fact, two-thirds of the book covers exploited marine species, particularly fish. In the last two chapters, Daniel expounds and expands on his work from the previous ten years on fish growth, coming back to his favorite themes and tweaking a few noses along the way: "A sojourn of more than a few months on any tropical coast should convince even a myopic Northerner that seasonality exists and is important also in tropical oceans."

Continuing along these lines, Daniel also discusses the seasonal growth of tropical fish and the growth marks on their otoliths. He notes that "working in the tropics . . . forces one to seek much wider patterns and their possible causes" and relays the comments of an anonymous reviewer from a northern country on some of his work: "Rubbish, may apply in the tropics, but not here." He takes the opportunity to remind the reader that all extant forms of marine life evolved in conditions comparable to those of our current tropical zones, and only colonized the temperate and polar regions later on. Besides their studies on fish, Daniel and his colleagues also used ELEFAN to analyze the growth of a number of marine invertebrates, including sea urchins, shrimp, shellfish, squid, and later, even jellyfish. In the final chapter of Longhurst and Pauly, therefore, Daniel chose to focus on these marine invertebrates, noting in his preamble that "somewhat like the mythology surrounding tropical fish growth rates, it is easy to believe that the basic biology of invertebrates is so different from fish that population analysis techniques relevant to fish must be inappropriate for invertebrates." Daniel returned to this debate years later in an article entitled "Why squid, though not fish, may be better understood by pretending they are."[10]

Between developing multiple versions of ELEFAN, training sessions for the FAO, the San Miguel Bay project, trips to Peru, co-writing a book with Longhurst, his habilitation to direct research, the other forty or so conference presentations he gave in twenty-two countries on six continents, and directing a dozen or so master's theses in the

Philippines, Daniel's feet barely touched the ground throughout the 1980s. "The number of items he has left in airplanes—books, wallet, passport, clothes—are enough to start a whole new persona," jokes Jay Maclean. "He even lost one of ICLARM's first portable computers, launching a diplomatic crisis with the administration. And another time, he called from somewhere in Papua New Guinea having simply used up all his cash, credit cards not being of much use there, and had to wait days for a money transfer from Manila. The result was that everyone around him had to fuss to some extent to make sure he was complete in his documentation, traveler's checks, and so on, before he left on a trip."

"I get dizzy when I think about those years," admits Daniel, who published nearly two hundred scientific articles, one hundred and forty of which he signed as first author, and who would often be absent for eight months out of the year. The stories he read to his children arrived via airmail, recorded on cassette tapes. Ilya and Angela don't remember suffering from their father's absences, though Angela admits, "This family wouldn't exist without my mother."

During the short periods he spent in Manila, Daniel followed a strict routine. Up with the sun, around five or six in the morning, lots of coffee and fruit on the terrace with Sandra and the children, an hour of traffic, either alone with his driver or with a few colleagues for an early-morning chat. He would also go to lunch with Jay, Roger, Deng, and a few others, almost always at the Singapura, a stone's throw from ICLARM. "They made an excellent Hainanese chicken and rice; I would order that or squid in hot sauce," reminisces Daniel, whom everyone describes as having "a healthy appetite." "He was famous for sampling the plates around him after he finished," Jay would write later. "And he usually finished first even though eating hardly slowed down his conversation. Then he might realize only at the end of a meal that he had no cash and would turn to the nearest person ('Oh my gosh,' as if it hadn't ever happened before) and grab his or her upper arm in a rough sort of plea for a loan, and we knew that he was quite sincere, just absentminded, and the lunch money thing actually became a light topic in itself about whose turn it was to pay for Daniel. Then, the next time, he would pay for everybody, and the whole thing would become confused, although we knew it was contrived by then and all took part because it was a way of decompressing after the inevitably weighty discussions we had during lunch, for Daniel's presence always meant that we would descend bathyspherically deep into whatever topic came up, whether science, politics, religion, or problems in the office, this before his soup even arrived."

In the evenings, Daniel would spend another hour in traffic to make the six-mile trip home, eat dinner with the family while talking about his research or current events, and go to sleep with the sun. "Daniel was no night owl and disappeared pretty quickly, leaving Sandra to take care of any guests they might have," Siebren Venema recalls. "My father burns the candle at both ends," explains his son, Ilya. "When he feels good, it gives him the strength to work hard, and when he's feeling bad, he hides behind his work—I can remember him at the house, running toward his office to escape."

The engine soon overheated, however—too many manuscripts and grant applications, too many nights in airplanes and too much jet lag. One day at the Singapura, without warning, Daniel Pauly collapsed face-first into his plate. "He'd been well-fed for a few years by that point, and it wasn't easy to get him out the door to take him to the hospital," his colleagues remember. After a few days of rest, everything was fine, but the doctors told him it was a serious warning sign. Daniel ignored it.

NATURE IN A BOX

"I T WAS LIKE a constant adrenaline rush—we knew we could die any second—it was the climax of our struggle, and all our secret activities were suddenly out in the open." February 22, 1986: The People Power Revolution was under way in the Philippines. Early that morning, Deng called Daniel. "I won't be coming into the office today. Don't ask me where I'm going," she said. She joined the human chain that was blocking the entrance to a military camp. "Part of Marcos's army had defected, so we went to protect the soldiers with our bodies, because they'd had the courage to say no. We lasted four days, and in the end, the whole army abandoned Marcos, and he fled."

The first peaceful revolution in the Philippines was a success: Corazon Aquino took power, the first woman president in all of Asia. "I get teary-eyed just thinking about it," Jay Maclean tells me. "My wife, Margie, was up on the front lines, helping out with the election and then going up to stand in front of the tanks. It was very emotional, scary, the prospect of bombs and tanks firing on you." This popular uprising meant the end of a long struggle, one that had been exacerbated by the assassination of Ninoy Aquino on August 21, 1983. "The Philippines has a martyr," one French newspaper announced. A longtime opposition leader who was imprisoned, then later exiled to the United States, Ninoy braved great danger by returning to his home country on a commercial China Airlines flight, traveling with journalists and television cameras. When they arrived in Manila, Marcos's thugs handcuffed him, hauled him off the plane, slammed the door shut on the other passengers and the media, made Ninoy get down on his knees, and put a bullet in his head before throwing his body onto the tarmac from the jet bridge. His wife, Corazon, left her life as a proud house-wife behind to take up her husband's torch and challenge Marcos's reelection in early 1986, causing the dictator's downfall and making her the shining light of a popular uprising in which thousands of Fili-pinos took to the streets.

"It was only then that Daniel, like a lot of people I knew, under-stood how involved I was," remarks Deng, who had worked with him for four years at that point. "When I came back, he opened up to me, talking about his life and what a dilemma it was for him to be a leftist working in an international research center with an expatriate's salary. But we agreed that it was through our work that we were going to change things," she concludes.

And Daniel did work, even under fire. His friends and family recall that, from his rooftop terrace, he would watch the dance of military helicopters and fighter jets over the city, evaluating the risk of tak-ing his family through Manila's different neighborhoods. His system wasn't infallible, however. "During one of the military coups in the late eighties, the ICLARM board was meeting in the Mandarin Hotel," recalls Jay. "We went out to eat, and I remember saying, 'I don't think we should be here'—there were people shooting up and down the streets in Makati. But Daniel insisted we get back to the office and con-tinue the meeting through the chaos. He was the driving force."

Manila was a political and climatic hotbed, but the family escaped for a few weeks a year to vacation in California with Sandra's family. Their flight usually stopped in Hawaii, which they also visited, and Daniel, of course, went to meet the local researchers. Among them was Jeffrey Polovina—a tall, slim math whiz from the Honolulu Labo-ratory of the National Marine Fisheries Service. Jeff had joined this unit of the National Oceanic and Atmospheric Administration in 1978. In the early 1980s, his supervisor, Richard Shomura, asked him to cre-ate a mathematical model of an entire coral reef, the French Frigate Shoals, called Kānemilohaʻi in Hawaiian. The English name for the atoll, located in the middle of the North Pacific 560 miles northwest of Honolulu, honors the French explorer Jean-François de La Pérouse, who, while navigating the shoals in 1786, nearly lost two ships, the *Boussole* and the *Astrolabe*. La Pérouse had come from California and was en route for Macao, Korea, Sakhalin, and Kamchatka, before going to his eventual doom in the Solomon Islands of the South Pacific.

The French Frigate Shoals is now part of the largest marine protected area in the world, Papahānaumokuākea. The atoll measures eight miles in diameter and has a landing strip at its northern end that dates from the Second World War, as well as a marine ecology research station, both of which are situated only a few yards above the waves. These sites have to be evacuated regularly during storms and tsunamis. Over the decades, scientists have studied every aspect of the local ecosystem: the coral, the reef's tiny inhabitants, the fish in the lagoon, the ocean sharks, the multitude of marine birds who nest and feed in the area, a sizeable population of sea turtles, and the world's largest group of Hawaiian monk seals. In this faraway island paradise, each researcher tended to obsess over a single species or group of species, each so unique and extraordinary that no one had bothered to work on the big picture yet. The scientists knew that the coral ecosystems are one of the most incredibly productive in the world, that they are an oasis of life in the midst of the mostly clear and nutrient-poor tropical oceans, but they also knew that their coastal fisheries were largely in decline.

Jeff Polovina wanted to understand the big picture of how coral reef ecosystems function in order to help with their conservation. He was familiar with models developed by colleagues in Denmark (for the North Sea)[1] and Seattle (for the Bering Sea),[2] and went to see them. He returned deflated, certain that their highly complex simulations, which require a huge amount of very precise information from various sources, could never work for the coral reef ecosystem of the French Frigate Shoals, nor for other marine systems for which much less information is available.

But Jeff was pragmatic and creative. He simplified his colleagues' models and ended up with the following structure: rather than considering species individually, he grouped them together according to their role in the ecosystem, like different professional categories in human society. All the phytoplankton made up one functional group, the zooplankton another, and so on for the reef fish, the crabs and cray-fish, the carnivorous fish, the birds, and the marine mammals. Polovina drew a diagram that looked like a subway map: each station, marked by a box, represents one functional group, and the lines between the stations indicate who eats whom. From there, you need to know, for each box over a one-year period, the total mass of all the organisms in that functional group, the total mass of food ingested (which has been pulled from one or more of the other boxes), total weight gain for each member of the group, and the mass caught by humans (if the group is exploited by fisheries). At the NOAA laboratory in Honolulu, everyone lent a hand to provide the data needed to complete Polovina's giant puzzle. Once that was done, multiple equations connected the boxes to one another, calculating the total mass and the total production of the ecosystem in terms of biomass for the year in question.

In an ecosystem-wide snapshot like this one, balance is the main concern: there has to be enough plankton to feed the higher levels, from the reef grazers to large predators like sharks. At the base of the whole thing is energy from the sun, which feeds microscopic algae through photosynthesis. In order to test his model, Jeff Polovina worked with Marlin Atkinson and Richard Grigg, who measured coral respiration. Though it seems incredible, the two researchers man-aged to enclose whole heads of coral and measure the concentration of oxygen in the water around them over time. The more intense photosynthesis is, the higher the rate of oxygen production, and the more the coral grows and produces biomass that will feed the other inhabitants of the reef. Thanks to their elegant coral trapping opera-tion, Atkinson and Grigg were able to estimate the biomass produced by photosynthesis. Their results were similar to those predicted by Polovina, validating his model. "No one was more surprised than I," he wrote ten years later.[3] Jeff named his computational routine Eco-path. Once it was used to model an ecosystem, it could also predict the characteristics of any missing boxes. Ecopath thereby made it possible to estimate the biomass of fish "produced" by an ecosystem, and potentially made available to fisheries, each year.

Always modest, Jeff Polovina published his results without much fanfare: "I was busy elsewhere and I had doubts about Ecopath's acceptance, given some criticism that the model was overly simplistic," he says. Daniel certainly didn't think it was overly simplistic, though, and explained to Polovina that, on the contrary, a simplified system was exactly what everyone was looking for. "He told me that if I made Ecopath user-friendly and wrote a user's manual, he would see that it was applied around the world," Polovina recalls. "I did my part, and he certainly did his. For several years, the requests I received for Ecopath material often arrived in batches by country; thus, I could track the locations of Daniel's seminars on Ecopath."[4]

"Thanks to Ecopath, Daniel reached another level," comments Jay Maclean. Daniel had been dreaming about an ecosystem-wide approach since Sakumo, and since San Miguel Bay, he had been working to make it real by studying the writings of Danish scientists Erik Ursin and Per Sparre, who developed multispecies models for fish populations. This was also a logical follow-up to his previous work: the various versions of ELEFAN made it possible to quickly estimate the biomass of each group of species that made up Ecopath, the two routines fitting together like Russian dolls. Daniel mentioned Ecopath enthusiastically in the book he coauthored with Alan Longhurst on the ecology of tropical oceans and in the document he produced for his habilitation to direct research.

In fact, while Daniel was working on his habilitation in Kiel in 1984, Silvia Opitz came knocking on his office door. She had met him at the oceanographic institute in the 1970s. Since then, Silvia had spent several years in Brazil before returning to Germany via the Caribbean, where she had become fascinated with coral reefs. And now she was looking for a thesis topic. "I have something for you," Daniel said right away, pulling out Polovina's freshly written article. And so Silvia became the test pilot for Pauly's brand of Ecopath. The duo set their sights on the coral reef ecosystem of the British Virgin Islands because, as Daniel notes, "we knew John Randall had collected a huge pile of data on the fish there in the 1960s." Daniel sketched a network of boxes connected by lines, an Ecopath model suited to the coral reefs of the Caribbean. To feed this needy brainchild, which would later become the most detailed model ever constructed of a coral reef ecosystem, Silvia drew on Randall's publications, but also on the FAO's archives in Rome. "Cornelia Nauen worked there at the time, and she was a big help," Silvia remembers. "My husband (who had come to assist me) and I stayed with her." After her Roman holiday, Silvia, who was soon to be a mother, returned to Kiel. "Other than the fish, it was tough to find all the necessary information for the other organisms in the coral reef ecosystem," Silvia tells me. "But Daniel was persistent and infinitely creative, and I became an expert at extracting data from the scientific literature myself."

Silvia started a family while she was working on her thesis. She tells me about those golden years over lunch on a snowy day in Kiel in early 2016. Silvia has a charming Berliner accent, fast-flowing and low-pitched. "With Daniel, we worked like equals—there wasn't a big age difference, and he's always gotten along well with his female colleagues. He's nothing like the stereotype of the dried-up scientist. He's kind of out there, but besides being a great researcher, he's also a great human being, very warm and funny."

Once he'd finished his habilitation, Daniel made only short visits to Kiel. In a way, this worked out well for Silvia because their scientific conversations were so intense that she needed time to process everything. During her thesis, tons of documents and manuscripts shot back and forth between Kiel, Manila, and Daniel's many other destinations via fax or post. "The few times Daniel was actually in Kiel, he edited my thesis chapters at impossible speeds, advising me on the scientific content but also on the writing. His worries about research funding, his colleagues' jobs, etc., none of that entered the conversation—he stayed focused on the science." Silvia would have liked to focus on science, too, but she also had to make a living. "It's hard for women, even in Germany," she admits. After a rich but somewhat tortuous career, she is still an untenured associate researcher at Kiel's oceanographic institute.

During the second half of the 1980s, Daniel wasn't into playing games of power and scientific strategy like his mentor Gotthilf Hempel, who recruited armies of young colleagues by cultivating his political connections. Influenced by the feminist movement of the 1960s, Daniel was, however, perfectly aware of the lack of support for women in research. He published an essay on the subject, which began with some insightful remarks on the influence of researchers' sex on their interpretations of ecological phenomena.[5] In particular, Daniel relates that all the male scientists consulted on the subject believe that male sea lions actively form their harems, whereas the only woman* to have studied the behavior of these amiable creatures concluded that it was the *females* who chose to share a limited number of males because all they do is lie around consuming fish without contributing to the care of the young. As Daniel concludes, "The problems which man(?)kind faces are simply too big to lose one half of the world's potential scientists just because some people confuse sex (a biological fact) with gender." He cites a large body of literature, highlighting the scientific basis of his claims, but also talks about his daughter, Angela, whom he has always told, "Yes, women can and should become professional divers, scientists and airplane pilots." Daniel mentions that "Special encouragement is thus needed, such as the fellowship created exclusively for female scientists by the Canadian government." This short text, not even four pages long, would resonate strongly with many of Daniel's female colleagues all over the world. "I thought Daniel was kind of macho," comments Coleen Moloney, professor of oceanography at the University of Cape Town in South Africa, "so I was very surprised."

---

* Daniel doesn't mention her by name, but that woman is Patricia Majluf.

Daniel continued tinkering with the first version of Ecopath through the late 1980s. But it was only after Villy Christensen's arrival in Manila in early 1990 that all the ingredients came together. Villy was born in Skagen, in Denmark's far north, where the waters of the North Sea and the Baltic mix together in spectacular fashion. Both sides of his family had been fishermen for generations. Villy grew up in the port of Hirtshals, thirty miles southwest of Skagen. "It looked like this," he says, showing me an aerial photograph: the port full of hundreds of boats, including Villy's father's, which bears the registration number s211. The young Christensen didn't become a fisherman, though, and he was the first member of his family to go to college, studying biology and mathematics at Aarhus University, then marine biology in Copenhagen. Hired by the Danish Institute for Fisheries Research in 1980, he worked there for ten years, mainly on herring larval growth. In late 1989, the development arm of the Danish Ministry of Foreign Affairs offered to assign him to ICLARM. Villy was aware of the political situation in the Philippines, which was undergoing a series of attempted coups d'état—and the risks to his family—but he decided to accept: "I had met Daniel in 1989 in Kiel, then again in the Netherlands, and we had very interesting discussions—that was my main reason for going."

As soon as he arrived, Villy took up the frenetic rhythm of work at ICLARM. He began by reprogramming Ecopath. "Polovina's version was a minefield: if you tried to do anything unexpected, it all fell apart. That whole generation of models was unstable, and it wasn't unusual for them to just not work at all." Villy's priority was to make the new version of Ecopath more flexible, "so it could adapt to any ecosystem." Additionally, the computer language had to be easy to use by a large number of people. Daniel and Villy also wanted to use Ecopath to assess the condition of the ecosystems they studied, notably the trophic level of the fisheries therein. Trophic level refers to the position of a group of species on the food chain or pyramid. The trophic level of plants is set at one, two for herbivores, three for the herbivores' predators, et cetera. Classifying food chains by trophic level was suggested in 1942 by Raymond Lindeman, but the idea was later rejected because species, or groups of species, often feed at multiple trophic levels. This is the case with omnivores, including us, *Homo sapiens*: we would have a trophic level of two if we were all vegetarian and be classified as level six superpredators if we consumed only marine mammals. Three decades later, William Odum and Eric Heald suggested recalculating trophic levels not as whole numbers, but as decimals.[6] Humanity as a whole therefore has a trophic level of about 2.21,[7] similar to the anchoveta, but with strong disparities between countries: in Burundi, where the diet is 96.7 percent plant-based, the mean trophic level is around 2.04, whereas Icelanders, whose diet is 50 percent meat and fish, have a mean trophic level of 2.57.

Daniel and Villy added a model to Ecopath that made it possible to calculate the mean trophic level of each species and the trophic level of the whole ecosystem. The duo also drew heavily on the work of Robert Ulanowicz to come up with an indicator for ecosystem health. "Daniel was fascinated by Ulanowicz's work," Villy recalls, "and especially by his 1986 book on growth and development,[8] despite how tough it was to get through." At the time, Ulanowicz was a professor and ecologist at the University of Maryland, where he studied the organization of flows of matter and energy within ecosystems. He developed a measure of ecosystem performance called "ascendency," which combines the diversity of groups of species present in an ecosystem, the degree of specialization of those groups, the intensity of the links between each group, and the speed with which energy and matter flow through the ecosystem. An intact natural system will have higher ascendency than an ecologically degraded one. Much impressed by this theoretical leap forward, which was also being used in cognitive psychology and economics, Villy and Daniel included a measurement of ascendency in the new version of Ecopath. "A very productive combination," Villy comments soberly.

With these additions, Ecopath II was born. The open-access program was made available to the scientific community in July of 1990 and would also be the subject of an article coauthored by Christensen and Pauly,[9] the first in a long series and a great classic in the domain of ecosystem modeling. In the wake of their publication, the ICLARM researchers organized a conference with the International Council for the Exploration of the Sea (ICES) on Ecopath II in Copenhagen in October of 1990. "There were twenty or thirty posters describing different ecosystems—in just a few months, the number of models available had doubled worldwide," recalls Daniel.

A photo shows Villy and Daniel—more than a little proud—standing on either side of one of those posters. The author is one A. D. Pongase of ICLARM. They often joked about him back in Manila. In fact, Pongase—roughly "put yourself there!" in Spanish—was a fictional colleague that Jay Maclean had invented while he was working in Australia. Jay had brought him along when he came to work at ICLARM, where Pongase signed reports and evaluations of scientific articles by colleagues abroad, especially when the work in question was lacking and the criticism particularly acerbic. The Copenhagen conference was the high point of the enigmatic Pongase's scientific career: he received a letter from the director of ICES congratulating him on his work. On a more serious note, the great Ulanowicz himself soon showered them with praise: "One day a parcel appeared in my mailbox containing some 50 or more quantified food webs, replete with accompanying ascendencies. It was perhaps the most startling and gratifying moment of my professional career... It may not be much of an exaggeration to say that the realm of ecosystems is being opened to us by Polovina, Pauly and Christensen through their 'ecoscope.' For that is what ECOPATH II and its associated analyses represent—a macroscope through which to view the structure and functioning of entire ecosystems."[10]

Villy and Daniel spent the next two years editing the scientific articles that resulted from the conference. Contributions came from every continent, covering lakes, rivers, aquaculture systems, coastal zones, coral reefs, and offshore areas. The final product, a four-hundred-page volume published in 1993, quickly made its mark on the whole research community in marine ecology.[11] When I complimented Villy on the work's scope, calling it the "Copenhagen conference proceedings," he stopped me right away: "No, they aren't 'proceedings'—it's a book, that's very important." Somewhat incongruously, the "black book," as Villy and Daniel call it (a reference to its dark cover), contains several models of aquatic ecosystems in France: Aydat Lake, the Garonne River, and the Étang de Thau. These three chapters are written by Deng Palomares, who is also responsible for modeling Lake Victoria, Lake Tanganyika, and Lake Chad. But what was Deng doing in France?

"Daniel was working with Jacques Moreau of the National School of Agronomy* in Toulouse," Deng tells me. "I wanted to do a thesis, and in 1988, I had been awarded a scholarship to study in Kiel—I had even started learning German at the Goethe-Institut in Manila. But just before I was supposed to leave, in spring of 1989, Daniel told me that Moreau was looking for doctoral students." Deng didn't have anything against the switch, and Daniel explained that Jacques Moreau knew and applied all his methods, so she wouldn't feel lost. Never short of arguments, he added that French was spoken much more widely than German, especially in Africa. "Four weeks later, I landed in France," Deng tells me, laughing. "I learned French in three months in Vichy. I liked the class—we also learned about French culture and cuisine."

In September, Jacques Moreau—a specialist in freshwater African fish and a somewhat whimsical personage, the very picture of a French colonialist—welcomed her to Toulouse. "Despite all that, Daniel was his god," Deng tells me. Indeed, Moreau later translated into French and edited a set of Daniel's method papers, covering everything from his thesis to Ecopath II.[12] Moreau was kind to Deng, if a bit odd. Her first week there, she was in student housing. Moreau came to see her, sat down on her bed, and asked for a cup of coffee. Deng answered, "*That* is my bed, *that* is a chair. My bed is for me, the chair is for visitors. If you want coffee, we can go to a café. I don't make coffee for anyone, not even Daniel." Laughing, she admits that, even in Manila, everyone made coffee for Daniel without his needing to ask, simply because he

---

* The National School of Agronomy, or École nationale supérieure agronomique (ENSA) in French, is a highly selective and prestigious teaching and research institution. (TN)

didn't have those kinds of sexist attitudes. Moreau's main job was to protect Deng from racist colleagues, particularly one Monsieur Gilles, a perpetual doctoral student who whiled away his time at the Toulouse laboratory mumbling about "these foreigners who take the French government's money." When he became violent, Moreau sheltered Deng in his office, but Gilles managed to sneak in and destroy the floppy disks on which Deng had saved her thesis manuscript. "Luckily, I had hidden copies everywhere! Moreau was in a fit, and Daniel called the student to tell him that he was going to break his face."

Deng's memories of her time in Toulouse aren't all sunshine and rainbows, but when it came to her research, everything went smoothly. "One of the goals of my thesis was to show that our methods of analysis for fish growth were just as valid for freshwater fish. It turned out that salinity doesn't make a difference, which confirmed the importance of temperature." As always, Deng was one step ahead. "When I arrived in Toulouse, my thesis was already well on its way. I had a hypothesis, the bibliography, the introduction, methodology, some of the derivations. All that was left were the experiments and the modeling. I spent a year measuring fish growth, and in eighteen months, it was done."[13] Deng used Ecopath, with which she had become familiar in Manila. "At ICLARM, I worked with Astrid Jarre, who was good with numbers. She put together a series of models of the Peruvian upwelling ecosystem." Deng, for her part, put all of Moreau's beloved African lakes "in a box" before doing the same with the Étang de Thau and other French aquatic systems. She especially liked Auvergne and its Aydat Lake, as well as Lake Pavin, which was modeled by her colleagues in Clermont-Ferrand. Impressed by her work and her mastery of French, the Clermontois team offered her a teaching position at Blaise Pascal University—it was a tempting offer, but she still felt that going home was the right thing to do. She politely declined.

ECOPATH II MADE a splash in the 1990s, but it was still a static model, a snapshot of an ecosystem's condition, whereas ecological processes are by nature constantly in motion. In order to create something more dynamic, Astrid Jarre, another of Daniel's doctoral students, generated separate Ecopath simulations of the Peruvian upwelling system for each of the twelve months of the year—but it was hardly a perfect fix.

The next qualitative advance would be initiated by Carl Walters. I meet Carl for lunch in Vancouver in 2015. He seems happy with his status as emeritus professor at the University of British Columbia, which allows him to take his afternoons off to do yoga and visit with his granddaughter, whose photo he shows us proudly while we eat our sandwiches. A half day is more than enough for him to get everything done—especially since he wakes up at five in the morning. We have an interesting chat about animal physiology and then about the world economy. I like Carl a lot and I know he's a force to be reckoned with in mathematics, the laureate of half a dozen big international prizes for his work. A native of Albuquerque, New Mexico, he studied at Colorado State University and defended his thesis there in 1969. At just twenty-five years old, he immediately found a job at UBC, where he would work his whole career while also consulting for the Canadian Department of Fisheries and Oceans.

When Daniel met him in the 1990s, Carl had recently published, among a slew of other scientific articles, a huge tome on the adaptive management of renewable resources,[14] and an equally hefty volume on fishery management, coauthored with Ray Hilborn.[15] In this latest book, he had forcefully criticized ecosystem models as too unreliable to be of any use. And yet, when Polovina, Pauly, and Christensen launched their first Ecopath training session in Vancouver in 1995, Carl, always curious, joined in quietly alongside the students. As the days passed, his visits became more frequent until, one morning, he arrived with a spreadsheet he'd cobbled together overnight. Carl had transformed Polovina's linear equations into differential equations—Ecopath had acquired a temporal dimension, making it possible to simulate the dynamic evolution of ecosystems over time. Ecosim was born.[16] This computer program, improved over the years, would also make it possible to test the impact of all kinds of fisheries management schemes: "What will happen in my study area if fishing is reduced year after year, or if, on the other hand, it doubles? How many fish will be available if global warming reduces the growth of phytoplankton by half in a ten-year period?"

Once Ecosim went online in the form of an open-access program, Villy and his colleagues began spreading the good news, hosting more than fifty training sessions in twenty-two countries. Ecopath with Ecosim (EwE) would be used by six thousand researchers, inspiring a whole community to generate not one but a whole family of models, not unlike the different climate models used by scientists at the IPCC.* In 2017 alone, EwE was the basis for sixty scientific publications, the latest of which is currently spread across my desk: this article tells the story of how Catalan scientist Marta Coll's team collaborated with Villy Christensen to make a simulation of ecosystem dynamics in the Mediterranean Sea for the period between 1950 and 2011. Their work shows that there has been a strong downward trend in both the populations of fish and their nonhuman predators, as well as the cause of this decline: a combination of climate change and overfishing.[17]

---

* The United Nations' Intergovernmental Panel on Climate Change.

FOR ALL THE FISH
IN THE WORLD

K IEL, FEBRUARY 1992. I was studying for my final exams in physical oceanography and marine biology. There was a gargantuan amount of knowledge to absorb, from the cardiovascular system of the dogfish (*Scyliorhinus canicula*) to the dynamics of polar sea ice. Over the last six months, review sheets had piled up in my basement apartment, a sort of burrow that I shared with my girlfriend, Katharina, for whom I had left France three years earlier. Meanwhile, she toiled away at her studies in political science and art history—we rarely saw the light of day, sometimes holding contests to see whose skin was paler and washing down copious amounts of buttered toast with black tea. A virus I had caught during my annual visit home to France had affected the nerve endings of my inner ear, leaving me with six months of seasickness on dry land. At first, the doctors thought it was a brain tumor, but ultimately, a young tropical disease specialist identified the little vermin that was responsible for my suffering. No treatment was necessary; I just had to wait for my synapses to regenerate on their own. When I finally emerged from our subterranean shelter, I walked stiffly, forced to keep my gaze fixed on the horizon to avoid careening into the gutter—for years after that, I couldn't look up at the sky without falling over.

Time was running out to prepare for exams, so I kept cramming, with some help from my classmates, especially Antje Helms. We studied together once or twice a week in her spacious, sunny apartment in the town center. Sitting on the floor of her room, we would pick review sheets from a pile at random and the oceanography game show would begin. Antje (today a campaign manager for Greenpeace) always knew the right answer—my performance was a bit less consistent. I was grateful for her gentle encouragement, and her calm attitude helped me keep a level head.

Among other things, we were expected to have an encyclopedic knowledge of fisheries, and Antje convinced me to have a look at the new catalog of fish species available at IfM's library. I wove my way through the bookshelves and found myself in front of the library's only public computer, where I was given a few floppy disks with a neat logo featuring two fish, nose to nose: a bluish specimen from the African lakes and a pink coral reef dweller. The first floppy clicked and hummed in the disk drive, giving up its secrets. The interface was primitive, typical of the years before Windows took over. Since I was in a Hanseatic city on the Baltic Sea that, like Lübeck, owed its early fortunes to the herring, I decided to look up the page devoted to that species first. There was no photo at the top of the page, but rather a digital drawing of the silver fish, whose features had been reproduced pixel by pixel. Following the species' Latin name (*Clupea harengus*) were its names in various European languages, its biological characteristics, its distribution range, and many other interesting tidbits. This up-to-date summary would save me hours of library research, and I printed off the page enthusiastically, along with the ones for cod, smelt, and plaice, making the dot matrix printer click and hum for a good fifteen minutes. Like a few hundred other people in the year 1992, I had just discovered FishBase. A quarter century later, in 2017, this ichthyological encyclopedia would have millions of users all over the world.

"IN 1985, WHEN Daniel came to Kiel to get his habilitation to direct research, I had the office next to his. He was fascinating. It was like he'd switched a light on: he did a single presentation, and all of a sudden, I understood why I was studying fish growth, mortality, the reason for it all. Plus, Daniel was also way cooler than my professors at the time.

There were a lot of pacifist protests going on against the Pershing missiles, and he wore the badge with the white dove on the blue background."

Rainer Froese tells me about his first meeting with Daniel, in the hall of the oceanographic institute in Kiel, where he still works. As his name suggests, Rainer Froese is from Northern Germany. Rainer also has the dry accent and straight spine characteristic of the men from that region, softened by several decades of research in the tropics. A native of Wismar, in the former East Germany, he grew up in the Rhineland after his parents fled the communist regime. His father had served in the Wehrmacht during World War II, so when Rainer received his high school diploma in 1970, the young man was less than eager to complete his mandatory military service. He went to sea to escape conscription, earning his commercial captain's license and navigating the oceans for two years, mainly aboard supertankers. "I understood that sea travel, especially tankers, was polluting the oceans, that the whole enterprise was only about halfway legal," he says. Rainer was first mate and radio officer, so he often received messages from the shipowner, such as, "Make sure the hold is clean when you get to Rotterdam"—a thinly disguised order to pump seawater into the empty tanks, then release the water and oil residues back into the ocean, an illegal practice. "I decided to quit my job as first mate, even though it paid well, and start from scratch as a biology student, with the goal of getting food from the oceans instead of polluting them."

Rainer moved to Hamburg, then Kiel in 1980, where he completed his master's degree and his doctorate. A lifelong aquarist, he financed his whole college education by selling aquarium fish—"It was a crappy job; I was happy to turn it back into a hobby after that," he tells me, laughing into his beard. In 1987, Daniel was passing through Kiel again when he learned that Rainer had an interest in artificial intelligence and expert systems.* "Those were the early days of computing," Rainer

---

* Computer programs using some form of artificial intelligence to simulate the assessments of people with specialized knowledge, e.g., in medicine or taxonomy.

remembers. "I had started programming, and I designed a routine that could identify the species of fish larvae based on their morphological characteristics. Daniel burst into my office with boxes of index cards, each one containing growth parameters for a species of fish. It was the database he'd created while working on his thesis, with more than five hundred species. Daniel wanted to make it available to his colleagues in developing countries." The man himself would give a more detailed description of his motivations in the introduction to a 1991 article on the subject, coauthored with Rainer:

> "Unfortunately, nothing is known on the biology of..." How many times have we read this silly little phrase—or a variant thereof—in papers or reports on the resources forming the basis of tropical and subtropical fisheries? A silly phrase it is because it is generally not true—it reflects only the information available to its author.[1]

Daniel was raring to go as usual, but Rainer puzzled over the best way to move forward: "Daniel's cards looked simple, but to put them in an electronic format, we'd have to create twenty, thirty different tables that all had to be connected to each other." With a draft of the database in his luggage, Rainer came to Manila in 1988, where he stayed with the Paulys for two months while he talked Daniel and his colleagues through the exact content of what would come to be called FishBase: species classification, morphology and physiology, population dynamics, ecology, reproduction, diet, et cetera.

Daniel quickly presented the project to ICLARM, announcing that the database would cover some 2,500 species. "There was a moment," Rainer remembers, "when I was sitting in this tiny office staring at my computer and I realized that if we started this, it would eventually have to include *all* the fish in the world, that it would become this huge thing." Indeed, at the time there were already over twenty thousand fish species known to science—but nothing could stop Daniel now. Rainer insisted on setting up a collaboration with the FAO, where Walter Fischer and his successors had already created similar databases starting in the 1970s. Daniel dragged his feet: "They'll only create more problems . . ." But the two eventually flew to Rome early in 1989 to sign a pact of nonaggression with their UN colleagues. "After that, one of our partners from the FAO worked on the project for a year, then he left, and from that point on we didn't get anything else out of them," Rainer sums up.

At ICLARM, on the other hand, it was only the beginning. Daniel and Rainer bought one of the first PCs equipped with the famous Intel 80386 processor and hired Susan Luna and Belen Acosta, who would be in charge of inputting the data extracted from every available scientific resource: books, articles, reports . . . Then Rainer submitted a grant application to the European Commission, which was approved after much internal lobbying by Cornelia Nauen, Daniel's lifelong friend and former classmate, who had already been working at the EC's Directorate-General for International Cooperation and Development in Brussels for several years. Rainer moved to Manila with his family in 1990 and stayed for ten years. The FishBase team recruited like crazy, and the number of PCs multiplied, the machines lined up and transformed into a digitization chain. They passed the six-thousand-species mark in 1992; then in 1994, the floppy disk gave way to the CD, which they distributed to four hundred research organizations in sixty-two countries. In 1995, Robert McCall and Robert May of Oxford University test-drove FishBase for the scientific journal *Nature*. In an article titled "More Than a Seafood Platter," they conclude:

> In short, *Fishbase* draws together and makes accessible a huge amount of information about fish and fisheries, much of which was previously buried in the "grey literature" of reports from fisheries institutes or working parties . . . Perhaps most important, and certainly closest to the authors' hearts, it will benefit developing countries, where the lack of comprehensive libraries is often keenly felt.[2]

In 1996, FishBase went on the internet, a technological advancement that couldn't have arrived at a better time for Rainer, Daniel, and the rest of the FishBase team, who had pioneered big data* and open access. The new home page for the piscine database, which now covered more than fifteen thousand species and had at the time already thousands of users, was the work of Tom Froese, Rainer's teenage son, who knew his way around computers. At the same time, a second, larger grant from the European Union came through, allowing the team to spread the good news throughout Africa, the Caribbean, and Oceania through training courses in Trinidad and Tobago, Kenya, Namibia, New Caledonia, and Senegal, where they promoted the use of FishBase and the orderly collection of information on marine biodiversity. The tireless Michael Vakily (who had been in ICLARM's orbit since the San Miguel Bay project) and Deng Palomares hit the road.

"We were a flying circus," Deng remembers. "I spent years up in the air—and I never want to travel like that again. One time, I ended up on a long-haul flight on Christmas Eve! Luckily, Michael is a sweet guy and I liked traveling with him. He's very humble and very positive, whatever happens. In Africa, for instance, we often had big money problems, and the administration was slow to respond, but Michael always kept his cool. He's one of the German colleagues who played an important role in Daniel's career, like we did in the Philippines." On the other hand, at the beginning of their collaboration, Deng found Rainer Froese a little too serious, but she learned to like him. "Rainer prefers objectivity to optimism, though he'll respect you even if he disagrees with you; he's someone who has a lot of heart even if he doesn't show it, and I trust him."

FishBase's stated objective was to cover every fish species by the year 2000, the equivalent of a forty-volume encyclopedia. The team now had fifteen members, including eight working on encoding data.

During my stay in the Philippines in 2017, I met most of the original FishBase team. The team, which is mostly made up of women, is now based in Los Baños, on the compound of the International Rice Research Institute (IRRI), part of a huge campus belonging to

---

* The use of extremely large datasets, a practice that is now very widespread.

the University of the Philippines, where the heat and humidity are slowly moldering the buildings into a state of romantic dilapidation beneath the shade of some giant acacia trees. Their offices look out over huge experimental rice fields and, farther away, the last remains of the old-growth forest that covers Mount Makiling, where hikers brave bloodsucking leeches. The whole FishBase team heaved a sigh of relief in 2000 when the project left behind the pandemonium of Manila to set up shop in this green paradise.

In the early 1990s, FishBase recruited its staff from among Daniel's former students at the University of the Philippines, notably its Marine Science Institute. Apparently, he'd made quite an impression. "I like the practical aspect of his teaching," Emily Capuli tells me, adding with a laugh: "He really knew how to explain things in simple terms and he gestured a lot while lecturing—it was hard to fall asleep in his classes." Daniel, whose primary concern was academic excellence, appearances be damned, stood out from the very hierarchy-conscious Filipino professors. "He was very generous with praise," says Cris Binohlan, "which really boosted our confidence."

They all agree, however, that they were initially intimidated, even terrified, by what they perceived as a human mountain, a creature from another world. "His idea of a 'break' was to wander around the office and very obviously look over your shoulder. He even did it to people who didn't work for him," Cris recalls. "And since he was usually in his socks, we couldn't hear him coming, which was freaky." Daniel usually arrived at FishBase central in the middle of the day since their offices were on a different floor from his. "We were all on our computers, and he would come to see each one of us to ask us about our work, correct mistakes, teach us things," Armi Torres tells me. She adds, "In the early days, he would do a presentation on a family of fish about once a week, covering its population dynamics, of course. Later on, it was our turn and we would present the group for which we were synthesizing infor-mation at the time. Doctor Pauly is such a hard worker that we would have been ashamed of ourselves if we didn't do as much." Fishing for information in an ocean of not-always-accessible scientific literature written in several languages doesn't sound like a very exciting job to me. I ask the women about why they keep at this dry task, cataloging the fish of the world, day after day, year after year. "Some people aren't suited for it," Susan Luna answers calmly, "but I love it, and I never get tired of it. You need to have the soul of a lace-maker—actually, in our spare time, us FishBase girls organize sewing and knitting workshops."

Daniel and his young colleagues didn't associate much outside of work, but they got to know each other over the years. "After a while, we were close enough to be able to tease each other in a friendly way, without offending," says Emily Capuli. "When Holy Week rolled around, he'd say, 'When are you going to crucify your god again?' He was provocative like that, but it really made you think, and we could always agree to disagree." Cris Binohlan concludes, "He made jokes, we joked back. I would tell him, 'You and Saint Peter have something in common—you're both interested in fishing!' He laughed till he cried!" When it comes to Rainer, the women all agree: he is the perfect boss— easy to talk to, relaxed, and good at giving clear instructions. "In twenty-five years, I think I've seen him lose his temper once," Cris tells me.

Classifying different fish species within FishBase soon turned out to be an intimidatingly complex task. Daniel and Rainer, who were no taxonomists,* sought help from a specialist, recruiting Nicolas Bailly of the Muséum national d'Histoire naturelle in Paris in 1996. "The first FishBase was set up as though fish taxonomy were complete, like there was nothing left to discover," Nicolas tells me when we see each other in Manila in early 2017. He has just arrived from France, and, like me, he is suffering through his jet lag in silence. "It usually takes me two weeks to get back to sleeping normally," he tells me with a sigh. Nicolas is celebrating twenty years with the FishBase project—of which he has been scientific director since 2011—and sixteen years of marriage to one Deng Palomares. Besides Nicolas's phenomenal knowledge of systematics,† Deng appreciates his Gallic frankness and his sharp

---

* People who have the skills to describe, name, group, and classify living organisms.
† The science of classification.

Parisian accent. On his end, Nicolas has adjusted marvelously to the paramilitary order that Deng imposes on their various residences all over the world. The couple has also racked up hundreds of hours of diving in many beautiful coral reefs. Nicolas never tires of exploring them, outfitted in vintage equipment worthy of Jacques Cousteau, nor of playing the clown to make his wife smile. During this already-torrid January in the southern hemisphere, Deng offers to put me up for a few days in their apartment in downtown Manila. In the mornings and evenings, Nicolas plays "Wish You Were Here" on the guitar, and we enjoy some quiet moments together.

In the first few years after its creation, the species classification system in FishBase did not conform to any particular norm. Starting in 1998, it was aligned with the Catalog of Fishes put in place by Bill Eschmeyer of the California Academy of Sciences, which makes regular updates following species discoveries, extinctions, name changes, and corrections. As a "diplomatic" gesture, Nicolas wrote an essay to confirm that FishBase was not intended to replace the Catalog of Fishes but rather to help complete it.[3] "Some taxonomists turn up their noses at FishBase," Nicolas admits. "We keep putting all the information online for free that they want to sell in their identification books." Later, Nicolas and Deng translated the entirety of FishBase into French, mainly so it could be used in francophone Africa; this version launched during a conference in Dakar in 1999.

Aside from coordinating digitization activities and research related to FishBase, responding to repeated criticism became one of Nicolas, Deng, Rainer, and Daniel's most important jobs. According to its detractors, FishBase was riddled with mistakes, and providing summaries of extant information was not in itself a legitimate research activity.

Their response to the first attack could be summed up as: "Why don't you do it yourself if you're so smart?" More precisely, Deng and Nicolas reminded everyone that, from their perspective, nothing could replace manual data input, which is still the norm at FishBase after twenty-five years. Indeed, automated data-extraction techniques like those used by the Encyclopedia of Life* lead to much higher error rates. These rates were initially estimated at 5 percent for FishBase. In the context of such a huge mass of data, that means thousands of errors, but it's all about your point of view: scientists using FishBase for a large number of species in the context of a comparative study would shrug, whereas a specialist looking for a particular piece of information on a single species would find such mistakes intolerable.[4] And what's more, because it is online and participatory, FishBase works on the potluck model: its users are also its contributors, and are therefore responsible for suggesting additions and corrections. "We accept criticism gratefully, even if it's harsh," Armi Torres tells me with a smile. "Insults are rare, and we do get praise." "Thanks to constant updates," Daniel concludes, "the error rate is way down, lower even than most of the published works on fishes."

When it comes to the second issue, though, the problem is mostly a cultural one. "I've never heard an oceanographer or a meteorologist question the fact that global databases are part of their sciences," asserts Rainer Froese.[5] Indeed, people who look down on humble naturalists counting critters in jars are less quick to dismiss those who do the same with whole galaxies. Nevertheless, in light of climate change and our present planetwide biodiversity crisis, there are now few who would question the importance of large-scale ecological studies and the construction of databases to go with them. This "big data" approach, still a novelty when FishBase debuted, has since become the norm. To quash any possible doubt on this point, Kostantinos Stergiou and Athanassios Tsikliras of the Aristotle University of Thessaloniki confirmed that FishBase had been used in 653 publications between 1995 and 2006.[6]

The encyclopedia's success is such that, in 2005, its founders decided to go a step further and create SeaLifeBase, with the goal of widening the catalog to include all marine species. With the support of

---

* See eol.org.

the Oak Foundation,* six additional research assistants joined the team in Los Baños. The first pages of the new database went online in 2008, and less than ten years later, it covered over 74,000 species and received 1.5 million hits per month from more than four thousand regular users all over the world.

The successes of FishBase and SeaLifeBase were intoxicating, but they hid an increasingly precarious financial situation. It became clear in the late 1990s that ICLARM, which had suffered a series of funding-related crises, could not support the programs long-term. Their funding from the European Commission was substantial, but no one thought it would last. In 2000, FishBase finally covered every known fish species, and Rainer, Daniel, and their colleagues came out with a prodigious user's manual. The same year, a consortium was set up between ICLARM, the FAO, and various institutions in eight countries. But this collaboration didn't cover salaries for the FishBase team and, in practice, balancing the budget has turned out to be difficult. Each year around the same time, Rainer, Cornelia, Deng, and Daniel break out in a cold sweat. This was notably the case during my visit to Manila in 2017.

While Deng pelted potential funders with messages, Cornelia Nauen swooped in to help once again, along with Loida Corpus, one of Daniel's former students, who is based in Singapore and had friends in high places. Tensions mounted in the meeting room, but strangely, out in the offices, serenity reigned supreme. I shared a work space with two of Deng's assistants, who are barely twenty years old. One of the young women hummed to herself as she worked, sitting cross-legged in an armchair. Susan Luna, who began working here at that age in the 1980s, tells me that she and her eight colleagues have been there since FishBase first launched: "We're like a big family—it's nice growing old together."

---

* Philanthropic foundation headquartered in Geneva.

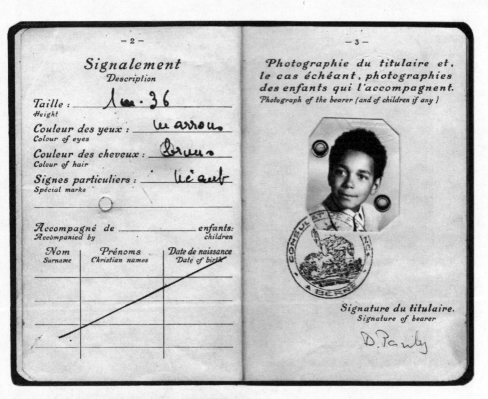

2

Signalement
Description

Taille : _1 m. 36_
Height

Couleur des yeux : _marron_
Colour of eyes

Couleur des cheveux : _bruns_
Colour of hair

Signes particuliers : _néant_
Spécial marks

◯

Accompagné de _____ enfants:
Accompanied by   children

Nom Surname	*Prénoms* Christian names	*Date de naissance* Date of birth

*Photographie du titulaire et,
le cas échéant, photographies
des enfants qui l'accompagnent.*
Photograph of the bearer (and of children if any)

Signature du titulaire.
Signature of bearer

D. Pan...

LEFT: La Chaux-de-Fonds, Switzerland, 1955. *"They always let me know that I was different. They told me over and over again that my mother had rejected me, that I should be grateful, that without them, I would have died of hunger. Little by little, their grief over losing their youngest son turned into greed, and I became their servant. As soon as I was old enough to work, I was seen as a resource."*

ABOVE: Avize, Marne, France, 1938. *She had her father's long face and her mother's steely gaze and sharp mind. Everyone agreed that Renée should continue her studies, and her schoolmistress supported her as far as she could. But her father wouldn't hear of it: "She'll work like the rest of them," he declared. And so at the age of fifteen, Renée began her first job as a domestic in Nancy.*

LEFT: La Chaux-de-Fonds, Switzerland, 1956. Daniel with
Monsieur G, *"a good man who was dominated by a terrible shrew."*

RIGHT: Wuppertal, Germany, 1964. *Daniel slept very little between
the factory, his classes, and his ever-accelerating bibliophagy.*

ABOVE: Daniel and Walter Kühhirt with Daniel's cousin Winifred Whitfield (Ed's sister) in California, 1969. *Meeting his family and seeing a Black America that was fighting for its civil rights was like an electric shock for Daniel's politics and identity.*

FACING TOP: Java Sea, Indonesia, 1975. *"A typical tow would yield two hundred kilograms (440 pounds) of fish, and there were 150 different species, of which only 80 were known," Daniel marvels.*

FACING BOTTOM: Baccalaureate, Wuppertal, Germany, 1969. *The article, from the spring of 1969, specifies that twenty-four of the twenty-five candidates, with an average age of twenty-five, passed the nine written tests and almost as many orals. Only a few of the candidates were interested in going on to higher education. A single student is named in the article, one Daniel Pauly from Paris, France, who declared that he had paid for his own schooling and wanted to go on to study biology.*

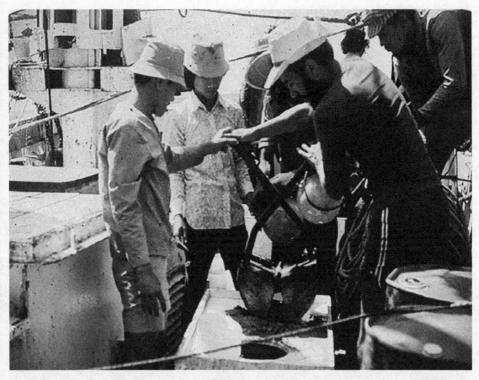

Mit Cicero, Algebra und Goethe
zur nachgeholten Studienreife

Von KLAUS-PETER HELMS

15 Absolventen dreieinhalbjährigen abendlicher Schulbesuche stellten sich am Dienstag und Mittwoch der Prüfungskommission unter Leitung von Oberschulrat Dr. Robert Walther, um das Abitur. Eintri-

Die Rasenn, die entgegen den Gepflogenheiten sonstiger abendgymnasialer Ausbildung diesmal am Tage vorgenommen wurden, brachten 18 junge Männer und zwei Mädchen zum Wissenzuwachs zusammen. Bei den Prüfungen war auf Wunsch des Düsseldorfer Schulkollegiums der Wuppertaler Schulausschuß vorgeschrittener Stufige Leiter des Abendgymnasiums, Oberstudienrat Otto Hoim, bereits zugegen.

Das Durchschnittsalter der Prüflinge lag bei 28 Jahren, also sind in der Schulordnung gemäß – bereits in Berufen tätig. Fortkommen im verantwortungsmäßigen Beruf, wie Wechsel in völlig neue Berufe waren die Beweggründe für die dreieinhalbjährigen Strapazen.

Die NRZ fragte den Vorsitzenden des Prüfungsausschusses, Oberschulrat Dr. Robert Walther, nach seinen aktuellen Problemen dieses Schultyps: „Natürlich 1958 als eine Erweiterung des Fächerkanons am Abendgymnasium dazu, aber jetzt in einem System, das mehreine monatigig läuft, muß wird es viel zu umständig, entsprechende Pädagogen etwa für Biologie oder Erdkunde und Biologie zu finden." Auf die Frage, ob das Abitur am Abendgymnasium mit leichtundem abgewehlten weniger ausmache „Botschaftsrat! Walther: „Der halte ich unverändentem Protestieren; der Fächerkanon ist so was zu beschänkkend...und es bleibt die Anspruch and Allgemeinbildung!"

Neun Studienfächer

Damals stießen mit dem Abendgymnasium durch neun Studienfächern hinabrücksichten, die alle verzüllich aber nicht mehr absicht, auch mündlich geprüft werden.

Die Absolventen sollten angesichts ab dieser „spitten Forderungen, „Ich habe einmal als Arbeit begonnen, in welche ich aus auch so

Ende fühng!", erklärte einer der Prüflinge, heute Journalist mit Ambitionen für das Wortblatt. „Ich brauche in diesem Beruf kein Studium", meint der Prüfling, der die Universitätsnote gereizt erringen könnte. „Wahrscheinlichkeit orientiert mich in der nächsten Zeit damit befassen, entschloßen sich das wider zu vorgraben, was mit für diesen Tag ausgeschöpft hatte...".

Auch eine Abschlußnote ist eigentlich noch mehr nötig. Das einzt die Examenwende, sowohl die bereits über die Begrinvorerörterprüfung ihre Erzählung hatte.

Viele Studienwünsche

Von Bedeutung dagegen war der erfolgreiche Abschluß für alle zukünftigen, Daniel Fauly, etwa, seit fünf Jahren in Wuppertal lebender Franzosen, gehörig in Paris, der auch nach eigenen Worten in der

vergangenen Jahren mit „Jure" durchgehandel hatte. Er will nun jetzt dem Biologie-Studium zuwenden.

Die Studienwünsche der Absolventen kreisten reichen vor der Germanistik zur Mathematik, von den Biologie bis zur Medizin. Wenige allerdings wollten noch nach der Weg in gehobener Laufbahnen des Beamtentums erschließen.

Offenbar in Berufen schon beweht, wollte ein paar der „Examens-Senioren" Lompedikulary vor ihr Prüfungskommission. Wie erklärt Prüfsäler diskutierten die nach der Prüfung via der Kommission über Mathematikaufgaben und ihre Lösungen. Nur wenige, zeigten sich geniessen. In dieser Fälle war noch längst der Prüfung auch ein Tauschwahre verbinden...

Heide Diskussionen in den Pausen zwischen den Prüfungen in den einzelnen Fächern. Gestern wurden die Abiturprüfungen beim Abendgymnasium abgeschlossen. Von 15 Teilnehmern bestanden 14.

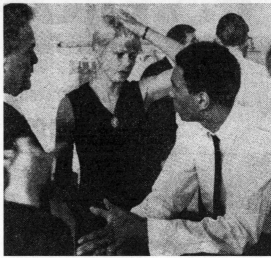

LEFT: Ed Whitfield, at the Presidential Scholars ceremony, the White House, Washington, DC, 1967. *"I wanted to refuse it. At the time, I was already active in the NAACP youth chapter and very much opposed to the war in Vietnam. And then I thought about it and I realized, if I do this, my mother will kill me. That's one thing that I kind of regret. I would have been the only person ever to do that."*

RIGHT: Ed Whitfield, Greensboro, North Carolina, 2016. *"My ambition in life was never to try and figure out if I could make myself successful. I was always trying to figure out how I could bring my community along, and that's why, fifty years later, I'm still marching, even if I'm not as quick as I used to be."*

PAULY JOINS ICLARM

ICLARM welcomes new Postdoctoral Fellow Dr. Daniel Pauly who joined the staff in early July. Concentrating on a subject with which he is already familiar—multispecies tropical fisheries—he will develop a program in cooperation with colleagues working in this field, leading to a theory of management of multispecies fishery resources, and test and apply the theory.

Dr. Pauly's experience with tropical fisheries suits him ideally to his present

TOP: Manila, the Philippines, 1979. *"By inviting a young Black scientist, Jack Marr showed that he had an attitude that was uncommon for a white American of his generation,"* Daniel notes, remembering that Marr welcomed him with a very simple to-do list: *"You are going to develop a theory of tropical fisheries."*

BOTTOM: Kiel, Germany, 1972. Daniel's room (1969–1974).

ABOVE: Kiel, Germany, 1978. "We decided that we were both 'young, gifted and Black' and we should get married. He asked me if I would, and I said, 'Yes, surely why not.' That's how that happened. It was just one of those very pragmatic things." She moved in with him in Kiel and never returned to London to finish her doctorate.

LEFT: Daniel with the Wade family (except for Sandra's sister, Cheryl, who took the photo), Seaside, California, 1980. "For a Black family, the United States was hostile territory—you had to plan your trip as if you were crossing the Sahara, identifying the oases in advance so that you could stop safely."

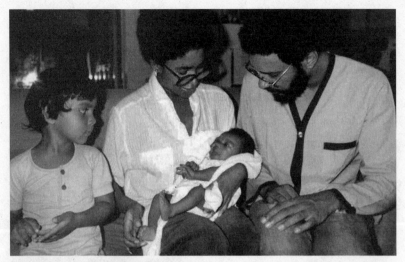

ABOVE: Manila, the Philippines, 1981–83. *The next year, Sandra gave birth to a little girl. The Paulys named her Angela as a tribute to the famous African American activist… A photo from that time shows the Paulys, now a family of four, standing under some palm trees on the beach: a peaceful postcard from their tropical exile.*

ABOVE: Daniel with Maria Lourdes Palomares, aka Deng (to his left) and Isabel Tsukayama (facing Deng), Callao, Peru, 1987. Deng: *"There was a real intellectual compatibility between [Isabel] and Daniel even though she was very conventional and he was a lot more rock and roll."*

LEFT: Deng Palomares, Manila, the Philippines, 1983. *Deng's story is emblematic of the destiny of many of Daniel Pauly's young colleagues from the tropics, unfairly caught up in the turmoil of postcolonial history, far from the comparatively insulated atmosphere of research laboratories in the Global North.*

LEFT: Kiel, Germany, 1984. *"I get dizzy when I think about those years,"* admits Daniel, *who published nearly two hundred scientific articles, one hundred and forty of which he signed as first author, and who would often be absent for eight months out of the year. The stories he read to his children arrived via airmail, recorded on cassette tapes.*

RIGHT: Daniel on the day of his examination for his habilitation to direct research, Kiel, Germany, 1985. *Suddenly without secure funding, Daniel returned to Kiel in 1984 with a contract pieced together by his allies at the oceanographic institute that gave him a few classes to teach and six months to obtain his habilitation to direct research... At the end of 1984, he went into his oral examination with the same confidence and found himself face to face with a whole jury full of bigwig researchers. According to his colleagues present that day, Daniel shone once again by his impertinence.*

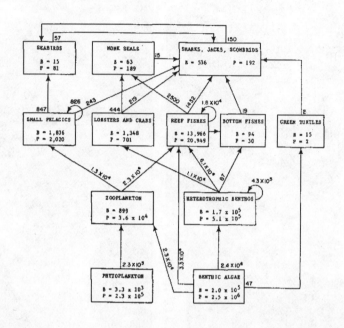

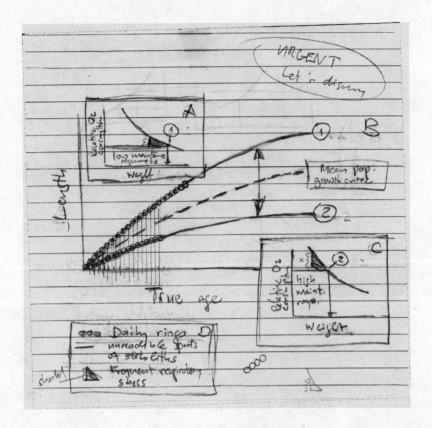

FACING TOP: Ecopath model, Honolulu, Hawaii, 1984. *Polovina drew a diagram that looked like a subway map: each station, marked by a box, represents one functional group, and the lines between the stations indicate who eats whom.*

FACING BOTTOM: Ecopath Conference, Copenhagen, Denmark, 1990. *A photo shows Villy and Daniel—more than a little proud—standing on each side of one of those posters. The author is one A. D. Pongase of ICLARM. They often joked about him back in Manila. In fact, Pongase—roughly "put yourself there!" in Spanish—was a fictional colleague that Jay Maclean had invented while he was working in Australia.*

ABOVE: Fish growth, Manila, the Philippines, 1990. Annabelle Cruz-Trinidad: *"Daniel used to give out really difficult tasks around five PM; the staff would dread that moment, but he would call us 'marvelous' in the most charming way."*

FACING TOP: Tubbataha, the Philippines, 1992. Jay: *"Daniel's diving followed the same absentmindedness as his terrestrial pursuits. His near-legendary exploits sometimes brought him into close proximity with 'Z'—the fisheries scientists' term for 'total mortality.'"*

FACING BOTTOM: Manila, the Philippines, Christmas 1984. Daniel: *"Before ICLARM had its crisis, I'd intended to spend the rest of my life in the Philippines, maybe retire to the seaside, in Anilao perhaps. I liked the Philippines, and the feeling was mutual."*

ABOVE LEFT: Winston McLemore, Daniel's biological father, Seaside, California, 1982. *Winston and his wife, Mary, welcomed Daniel into their home, but things didn't click the way they had with the Whitfields. Winston voted Republican and his worldview was diametrically opposed to that of his son.*

ABOVE RIGHT: Louis Pauly, aka Loulou, La Creuse, France, ca. 1994–95. *Louis had been raised by social services—after growing up in various orphanages, he had generous ideas about family ties, and he adopted Daniel without even meeting him.*

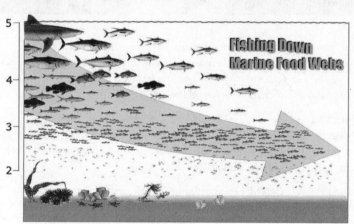

FACING TOP: Cornelia Nauen and Gotthilf Hempel, Kiel, Germany, 2016. *Cornelia had fled the Ruhr ("a concrete desert") and a stiflingly conservative home. Daniel's seriousness and maturity impressed her, and he appreciated her in turn for her absolute loyalty, rigor, and industrious good humor... At the oceanographic institute in Kiel, fisheries biology was a big deal, overseen by a major authority figure—Gotthilf Hempel. This formidable professor is the most influential German oceanographer of the twentieth century.*

FACING BOTTOM: Fishing down marine food webs, Manila, the Philippines, 1998. *Rachel Atanacio is the one who drew, pixel by pixel, each species of fish that appeared in the first version of that famous ichthyological encyclopedia. Illustrating fishing down marine food webs was her first time working with Daniel. "I'm an artist, not a scientist," she tells me. "He had to explain to me exactly what he wanted, but I suggested adding the diagonal arrow that highlights the general decline."*

ABOVE: Daniel and Philippe Cury at the award ceremony for the Albert I medal, Paris, 2016. *Philippe and Daniel have known each other "forever" but can't recall the exact date of their first meeting. It was almost certainly at a research laboratory in Monterey, California—the same town where Steinbeck set his novel* Cannery Row.

ABOVE: Jennifer Jacquet, New York, 2016. *According to Jacquet and her colleagues, the Marine Stewardship Council's criteria are quite simply too relaxed and the environmental benefits they promise too nebulous.*

FACING: Artisanal fishing in Senegal. *Over dinner, I was told about the heroic daily lives of the piroguiers and the scrappy resourcefulness of hungry children who snatch small fish off the boats as they come ashore. My African family also complained about the armada of foreign ships that were pillaging African waters. This shameful practice had been going on for some time already and took several forms, but to put it bluntly, the whole world was helping themselves to the marine resources off the Senegalese coast.*

FACING TOP: Renée and Angela at a retirement home near Guéret (in La Creuse, France), 2013. *"La Creuse helped me make my peace with France,"* reminisces Daniel, who enjoyed the beautiful countryside and talked with his mother from morning till night.

FACING BOTTOM: Ilya and Angela Pauly, Montreal, Canada, 2002. *"He's trying to wake people up, so I can forgive him for a lot, for being away so much, just because of that."*

ABOVE: The Pauly siblings, Louis Pauly's funeral, La Vallade, La Creuse, France, 2003. *"They were exhausted from all those years of work,"* one of their neighbors told me. *"Loulou didn't live to be very old. You could tell they hadn't been very happy in Paris, but here, they were."*

ABOVE: Daniel and his books, Vancouver, 2005. *"Of course, Darwin's ideas fascinated me,"* Daniel tells me, *"but as I read on, I got to know him as an extraordinarily sensitive and likable human being. That's not often the way with famous men—usually, when you get beneath the surface, you discover a monster."*

FACING TOP: Anilao, the Philippines, 1986. *He would, however, look down his nose at the joys of coral reef diving for many years. Once, he gloomily told Jay that "all I see down there is conflict, fish spending their time chasing each other and fighting—I'm swimming through the tragedies of the deep."*

FACING BOTTOM: Award ceremony for the Cosmos Prize, Tokyo, 2005. *Starting in the early 2000s, the honors just kept flooding in, making him the most decorated marine biologist in the history of the discipline.*

花の万博記念
「コスモス国際賞」
International Cosmos Prize
for 2005

ABOVE: Proust Questionnaire, *Nature*, 2003.
Q: How would you like to be remembered?
A: *As the one who showed that the effect of fisheries on marine life is equivalent to that of a large meteor strike on terrestrial life.*

III

ON THE WORLD STAGE

THE BIG LEAGUES

IN 1992, EVERYTHING seemed to be going wrong at ICLARM. Like a small business bought up by a multinational corporation, the organization had just joined the Consultative Group for International Agricultural Research (CGIAR). This development agency, created in the 1970s, is supposed to be working toward a future "free of hunger" worldwide. In 2016, its annual budget was around 800 million dollars. The organization is now headquartered in Montpellier, France, and some of my French colleagues jokingly call it "CIGAR"—a reference to its hazy, opaque nature.

During the period after ICLARM became part of CGIAR, Jay Maclean was the interim director of the team in Manila, a job that turned into a two-and-a-half-year-long ordeal. "It was a very stressful time," he remembers sadly, "so stressful that, sitting at my desk, often my neck froze." Once the merger was complete, CGIAR installed a new administrative director, imported straight from Canada, whom Daniel described to me as "an incompetent chauvinist, a real Donald Trump." Events rapidly confirmed this first impression after one of Daniel's assistants was the victim of an attempted rape at work—the suspect's identity was known, and Daniel immediately reported the incident to the director. "The new boss laughed it off, saying it wasn't a big deal." Very sensitive about women's rights, Daniel called an emergency meeting of ICLARM's senior staff, and they decided to act. The mutineers soon wrote a request asking that the new director be fired for serious misconduct and sent it to ICLARM's board. "Instead of firing him, those idiots decided to stomp out the protest by sending us a

series of 'political commissars.'"* Roger Pullin received an official rep-
rimand; Daniel was removed from his position as program head and
demoted to the level of a simple researcher. But it was Clive Lightfoot,
the easygoing sustainable-aquaculture specialist, who really became
their scapegoat. He was let go based on "arbitrary accusations made
up after the fact," Daniel tells me.

Clive's wife, who grew up in a family of Canadian union organiz-
ers, was having none of it—she studied the Philippines' labor laws
and, with Deng's help, quickly discovered enough procedural flaws in
the firing to back up a formal complaint. "Clive and his family stayed
with us for two months," Daniel recalls, "and eventually the board had
to give in to the law—Clive came back to work. We won that battle,
but we lost the war that followed." Clive came through the ordeal a
wounded man, however. "One day, Clive walked into my office and
asked to sit down," Jay Maclean recalls. "He told me he wasn't feeling
well, that he should probably go to the hospital. So, I said, 'OK, I'll walk
over with you.' It was only about five hundred meters [not even half a
mile], but he told me, 'I don't think I can get up!' I called an ambulance
and we had to take him down the steps on a stretcher. When we got
him to the emergency room, he started to vomit blood everywhere. I
had never seen anything like it. They had a meter going and suddenly it
stopped and he stopped breathing. Terrible time . . ." In fact, Clive was
suffering from a perforated ulcer, but he pulled through.

A new administrative director came in to replace the old one. She
traumatized the entire ICLARM staff, who, decades later, still lower their
voices as they tell me about "the Australian lady boss—the one who
worried about the color of the Post-its in our files, not the content
at all!" Around that same time, according to Deng, "CGIAR was infiltrated by CIA agents whose mission was to root out communists." For
Jay, it became clear that "they wanted to change us from a research
to a resource institute . . . a reflection of the Australian public service,

---

* A facetious reference to a type of political officer responsible for ensuring the
ideological purity of the Soviet armed forces. (TN)

which I knew all too well." Roger Pullin adds: "After the change, it was never about results anymore, it was about form, management. The 1980s were good—the organization trusted us to get funding for our ideas. But afterward, the scientists who could attract lots of money were replaced by liaison officers who would sometimes get the names of the countries where we ran our programs wrong!" Tensions ran high and terror reigned. "The regular employees," recalls Roger, "got so scared they signed horrible affidavits about their managers." Daniel was even accused of "racism" during that surreal time in ICLARM's history.

An audio recording survives from the period, captured by Roger during a meeting of the organization's management committee. "I knew it would be ugly, so I taped it with a portable tape recorder without concealing what I was doing." Among some other informative tidbits, Rainer Froese can be heard speaking up to ask the management committee if they realize what they are doing—destroying a winning team.

As for "the Australian lady," Daniel tells me he "knew right away that she was a liar." As much became clear during a philosophical discussion about truth. She referenced a classic Japanese film (*Rashomon*, the story of a murder told from four different points of view*), which, according to her, demonstrated that truth is subjective and essentially inaccessible. For Daniel, on the other hand, "that film shows that truth *is* accessible, but only if you aren't afraid of it." The conversation left him demoralized. He describes the events that followed in a few short phrases: "The first two or three weeks, she seemed friendly enough, but after that, the higher-ups must have led her to understand that she was there to do away with us. I still had two years left on my contract, but I knew it was time to go—there would have almost certainly been an ugly surprise when it came time to renew." Daniel, whose international reputation was already fairly solid, had received a few offers over the years, namely from Duke University in North Carolina and

---

* Directed by Akira Kurosawa, *Rashomon* (1950) won an Oscar for best foreign film and a Golden Lion at the Venice Film Festival.

the University of Miami in Florida. But he refused to accept a job in the United States: "When I'm in the US, I always feel like I'm holding my breath." Europe was also too racist in Daniel's estimation to be a good place to raise his kids. During the ICLARM crisis, the University of British Columbia (UBC) asked him to help evaluate the newly created fisheries research center in Vancouver. He made his first appearance there in 1993, flying up from California, where the Paulys were spending their Christmas vacation with Sandra's family. "It was only a two-hour flight," shrugs Daniel, who takes planes the way most people take the bus. "I did a presentation and Tony Pitcher suggested that I apply for a position that he'd helped create. It was for a specialist in tropical fisheries who would also train students from developing countries."

Daniel wavered. "Before ICLARM had its crisis, I'd intended to spend the rest of my life in the Philippines, maybe retire to the seaside, in Anilao, perhaps. I liked the Philippines and the feeling was mutual." Jay also warned him that Vancouver was a town "where nothing ever happens," and which had been voted "the most racist city in Canada" by *National Geographic* in 1992. This was because of the colonialist attitudes held by a British population, which had been living there for a little over a century, toward First Nations peoples and immigrants from France, Japan, India, and most recently China. But Daniel was short on options, and, convinced he'd soon be axed from ICLARM, he accepted. His friends and colleagues weren't far behind, leaving one by one an organization that would soon sink into oblivion after having shone so brilliantly in the 1980s. Villy Christensen returned home to Denmark in 1994, saying, "I don't want to be the one to turn the lights out." Jay was let go in 1996, the year his son Marlon was born, and still struggles to make ends meet as a freelance writer. Roger and Rainer held on until 1999, but they too departed, one to work as a consultant on the impact of oil spills, the other to become a senior scientist at the oceanographic institute in Kiel. Even Deng, despite her loyalty to her homeland, followed Daniel to Vancouver in 2001, leaving behind a permanent contract and the possibility of promotion within CGIAR in favor of a new life in Canada. Explaining her drastic decision, she points out, "I left because in Manila, the head of the organization was no shining star. I want to work for people who want to do something real, not just work for a salary. So, I said to Daniel, 'I go where you go.'" ICLARM, for its part, ended up moving to Malaysia, leaving only the FishBase ladies in the Philippines under the long-distance management of Rainer, Deng, and Daniel.

It was the end of an era—probably of one of the happiest times in Daniel's life, thanks to the loyalty and good humor of the colleagues who so often became his friends, partly because he himself doesn't really exist outside of his work. They leaned on one another's shoulders and shared a lot of laughs. Roger, Jay, and Rainer, the musicians of the group, put together the "ICLARM Band" with Roger on vocals, Rainer on base, and Jay on the keyboard. I saw them perform for the first time in a home video from the 1990s: good-looking Roger, largely up to rock-and-roll standards; Jay, messing around with his instrument and cracking jokes; and Rainer, in the background but very much in control of the sound tech. They liked to play blues and reggae, with some Filipino pop mixed in. As the recorded concert gets into full swing, you can catch a glimpse of the electric trio's wives off to one side, a joyfully diverse group. For Daniel's going-away party, Roger and the ICLARM Band sang "No ICLARM, No Cry," their own unforgettable rendition of the Bob Marley song.

Going to work for UBC in 1994 added a brilliant line to Daniel's CV, which already provided glowing proof of his international renown. He'd also made the strategic decision to develop his teaching skills during the 1980s. In the end, though, it was his long list of publications that won over the hiring committee: at forty-seven years old, he had more than three hundred titles to his name—an incredible feat for a scientist of his generation. He upheld his reputation as a man who was "born to write" and found himself plucked from an obscure development agency in the middle of the Pacific, far from first-world academic circles, and thrust into one of the forty most prestigious universities on the planet.

According to Roger Pullin, "Vancouver was the best possible place for Daniel to end up." Daniel, however, was not ready to uproot his family from Manila, where his children were in school and his wife had a good job. He negotiated the right to work in Vancouver during the seven months of the Canadian school year and spend the rest of his time in Manila, near his family and ICLARM, where the dreaded "political commissar" (maybe not so bad after all?) tolerated his presence. "That's how I was able to finish my work in Manila, mostly as a consultant on the FishBase project," Daniel remarks matter-of-factly.

In Vancouver, instead of starting out as an associate professor waiting for tenure, Daniel was recruited as a full professor right off the bat. In fact, Canada turned out to be a little more civilized than he initially thought—the university had an affirmative action program for visible minorities designed to increase the diversity of its staff. "At first, it really weighed on me," admits Daniel, who didn't crow over his new position, "because that way of moving up in the world is risky. For me, though, it was another reason to keep working like crazy, so no one could say I wasn't worthy."

Thus began five decisive years in Daniel Pauly's career, much of which he would spend up in the air over the Pacific Ocean somewhere between Vancouver and Manila. In Canada, he lived on the UBC campus, a peninsula surrounded by the Strait of Georgia and what remains of the old-growth forest with its giant Douglas and grand firs, some of which are nearly 200 feet tall and more than four hundred years old. The average annual temperature there is 11°C (51°F), compared with 28°C (82°F) in Manila, and it rains half the time, sometimes for a month straight. Daniel had moved into a tall apartment building just a hop, skip, and a jump from the fisheries research center, which was housed in old, ramshackle huts. Totally immersed in his work as usual, but now far from his family and friends, Daniel gradually isolated himself: "It's a bad habit—I love silence, even more than I love music. So sometimes, I'd spend days, especially on the weekends, shut up at home where no noise could get to me. It was at that time that I started to get tinnitus." These symptoms were likely the result of stress caused by overwork and an existence that Daniel later described as "torn between two scientific worlds: that of theory and that of the application of science to problems in the real world." Though he quickly stipulated that "these tensions proved to be creative ones."[1]

IN MANILA, DANIEL spent time with scientists from other research centers associated with CGIAR, particularly agronomists from the International Rice Research Institute. He noticed that they didn't hesitate to analyze rice production on a global scale, whereas fisheries research was still largely a local or, at best, regional affair. With help from Villy Christensen, Daniel decided to "rectify this shortcoming by reviewing the state of, and potential for, catches of fisheries for the entire world," arguing that this was only logical "since fish were a globally traded commodity."[2]

Feeding the world—that was a serious occupation for agronomists, especially since the creation of the Club of Rome in 1968, which led to the publication of The Limits to Growth, a founding document of the political ecology movement. That same year, Paul and Anne Ehrlich of Stanford University published an explosive pamphlet entitled The Population Bomb to inform the public of the consequences of exponential human population growth. In the years after that, the couple attacked the problem of estimating how much of the world's natural resources were being consumed by humans. In 1986, they partnered with Peter Vitousek, also of Stanford, and Pamela Matson of NASA to estimate the proportion of energy produced by plants (using the sun's rays) that was being appropriated by humans worldwide.[3] The team reached the unsettling conclusion that during the latter half of the 1980s, humanity had consumed 35 to 45 percent of the products of photosynthesis. For the oceans, on the other hand, Vitousek and his colleagues estimated a much lower percentage (only 2.2 percent) and concluded that "human use of marine productivity is relatively small... It is unlikely to prove broadly catastrophic for oceanic ecosystems."[4]

Daniel and Villy were intrigued but skeptical. They agreed, however, that calculating the mass of phytoplankton needed to feed all the inhabitants of the world's fisheries would indeed be an excellent way to illustrate humanity's impact on marine resources. They began by using the Stanford team's calculations, considerably revising their methods along the way. For instance, Vitousek and his colleagues had based their work on an "average fish" with a trophic level of three, as if every fishery on the planet only captured a single species with a single, uniform diet. As we saw earlier, the trophic level of phytoplankton and algae is equal to one (as they are the foundation of the ecosystem), that of herbivores is equal to two, and predatory fish who eat herbivores have a trophic level of three. In reality, though, the trophic level of an exploited species can vary between 2.1 (in the case of mussels and other bivalves) and 4.2 (for bluefin tuna, swordfish, and other superpredators).

So Daniel and Villy set up a model of an exploited ocean which, instead of being a huge bathtub filled with a single fish species, consisted of six major types of ecosystems (coastal, ocean, coral reef, freshwater, et cetera), inhabited by thirty-nine "model" species of exploited marine organisms of various trophic levels. The duo also used forty-eight Ecopath-type models covering all the world's aquatic regions to recalculate the proportion of energy that moves up from one trophic level to the next. They concluded that the proportion of energy transferred is only 10 percent on average. This might seem incredible, but in fact, only about 10 percent of the energy contained in a population of phytoplankton is available to the fish that feed on them—the rest is lost in the transfer process. As a point of comparison, consider transformers that convert 220 volts from the electric grid into the 10 or 12 volts used by our computers, which then lose a huge portion of their electrical energy to heat, explaining why our devices tend to overheat.

Next, the Pauly-Christensen team dug into the FAO data concerning the annual catches of the world's fisheries, adding the volume of bycatch (organisms that are caught but then thrown back, because they don't correspond either to the needs of the market or the regulations in place). Based on these total catch figures, they estimated the quantity of phytoplankton necessary to feed the thirty-nine "model" exploited species they had identified. Their calculations indicate that, on average, 8 percent of the productivity of the world's oceans is now being absorbed by fisheries—almost four times as much as Vitousek and his team had estimated. Even more incredible is the fact that, though the environmental impact calculated by Pauly and Christensen in the open ocean was still fairly low (2 percent), it goes up to 35 percent in coastal areas. The study implied that the planet's fisheries, which had been flirting with the symbolic figure of 100 million metric tons caught annually, were unsustainable in the long term—precisely the opposite of what most experts claimed. They had effectively torn a hole in the persistent myth of infinite ocean resources.

Proud of the results, and rightly so, Daniel presented their research at the annual meeting of the British Ecological Society in Manchester in 1994, where it was received as "the most important and disturbing piece of information from the whole conference."[5] Daniel hit it off with Sir (now the late Lord) Robert May, one of the most influential figures of twentieth-century theoretical ecology, who was soon to be chief science advisor to the British government and president of the Royal Society in London. Since Sir Robert (who prefers to be called Bob) was part of the "establishment," Daniel started the conversation with a direct attack, which didn't so much as ruffle his British colleague's amused good humor—they got on splendidly after that, and Bob encouraged Daniel to submit his work to *Nature*, the gold standard in Old World scientific publishing. "Already done!" Daniel answered impishly. *Nature* would in fact evaluate the article and publish it after a few minor revisions on March 16, 1995. To top it all off, the journal commissioned and published a commentary by John Beddington, another distinguished British ecologist, who after dissecting Daniel and Villy's work failed to find any weak points.

At forty-nine years old, Daniel had published his first big hit in a first-class scientific journal. The article quickly became one of his top ten most influential publications. "I was mighty pleased when this contribution appeared and basked for a few days in the afterglow of colleagues' congratulations (such as you get when you first publish in either *Nature* or *Science*)," he says. This state of exaltation wouldn't last long, however—in the world of research as in the world at large, the higher your star rises, the more you expose yourself to criticism. "I was soon brought back to earth by a tersely worded message from the editor of *Nature*," Daniel wrote later on. His Canadian colleagues had written to the journal in London questioning some of Daniel and Villy's calculations relative to the all-important transfer of energy between trophic levels—criticism that required a response. Daniel made his arguments and defended their work inch by inch—fortunately, the study's major conclusions survived this trial by fire, and *Nature* did not publish the critique.*

A few years later, the ecologist Stuart Pimm, a specialist of the environmental crisis and mass extinctions, reviewed Daniel and Villy's calculations in his book *The World According to Pimm: A Scientist Audits the Earth* (a title that seems to denote a healthy ego). Although they used slightly different methods, Pimm came to the same conclusion as Daniel and Villy. "I can't tell you how happy we were to survive that audit," remarks Daniel. Stuart Pimm gave a slightly different explanation for his work: "Why did I repeat the calculations? I didn't believe them ... It is nothing personal. I do not believe my estimate(s) any more than I do theirs. Scientists are suspicious, critical nitpickers who are always ready to abuse their friends and colleagues verbally and in print over such differences."[6]

If Pauly and Christensen's analyses proved difficult reading for ordinary folk, their general conclusions were, to the contrary, quite revealing and accessible to the general public. For the first time in Daniel's career, the news media covered his work, alerted by a press release from the editors of *Nature* at the time of the article's publication. The information they presented would be read around the world, most notably in the British newspaper the *Guardian*, which presented

* The critique was nonetheless published in *Fisheries Oceanography* in 1995; see Pauly, *5 Easy Pieces*, 16–17.

the study's general conclusions and discussed the geopolitical consequences of a widespread shortage of marine resources. Of course, it's
always somebody else's fault, and the English journalist pointed a finger at the Spaniards with their "19,000 boats and 85,000 men," who
as a nation consume "40 kilograms [88 pounds] of fish per head per
year," concluding: "The Spaniards have emerged as the 'villains' of the
Atlantic war." Nevertheless, the reporter had the integrity to add that
the debate is complicated and that "the lesson from the plankton calculations published today is that this is a war nobody can win."

Daniel would learn from this same episode that once information
falls into the hands of the news media, scientists usually lose control
over how it is used—and to what end. For example, the Associated
Press published an article that presented the study correctly, but the
journalist, after consulting a scientist employed by the fishing industry, included the following quote by way of conclusion: "Here in the
United States, the more urgent issue is whether we can continue to
feed exploding populations of marine mammals, such as seals and sea
lions, which are competing for seafood with human beings."[7]

This first article in *Nature* and Daniel's compatibility with Bob May
would soon prove doubly advantageous: later in 1995, Daniel received
an email from Oxford, one of the first electronic messages ever to
arrive in his inbox. The dispatch came from Bob, then one of the
advisors to a journal called *Trends in Ecology and Evolution* (TREE),
which publishes syntheses and opinion pieces about the great scientific
questions of the hour in ecology and the evolutionary sciences. One
of the writers Bob had solicited for an article had thrown in the towel,
and he needed a stand-in, pronto. Daniel accepted his invitation, producing a page of text in a flash, the very same day.[8]

As he told it in his TED talk[9] many years later, this short column
allowed him to express a "tiny, little idea," something that could be
"explained in one minute."

In his 1995 text, Daniel begins by discussing the context in which
he will develop his "little idea." Fisheries are a global disaster for three

reasons: (1) Fishing fleets are heavily subsidized and two or three times larger than required to optimally exploit available resources. (2) Discarded bycatch, which isn't taken into account in the global catch figures (officially ninety million metric tons per year worldwide at the time), means that the actual catch is even larger. (3) Of the 260 fish stocks monitored by the FAO, the majority fall into the following three categories: collapsed, depleted, or in recovery. The solution is to manage global fisheries in a way that allows stocks to regenerate and grow large enough to support sustainable exploitation.

However, Daniel also identified a problem that has more to do with cognitive psychology than ecology:

> Each generation of fisheries scientists accepts as a baseline the stock size and species composition that occurred at the beginning of their careers and uses this to evaluate changes. When the next generation starts its career, the stocks have further declined, but it is the stocks at that time that serve as a new baseline. The result obviously is a gradual shift of the baseline, a gradual accommodation of the creeping disappearance of resource species.

Daniel calls this problem "the shifting baseline syndrome"—a little idea that quickly spread around the world and even made it into the dictionaries.

In fact, this concept applies to a great many environmental problems. For example, when he was ten years old, my father spent the summer of 1941 in the French seaside town of Palavas-les-Flots. When he tells the story of his three-month stay there, he describes wandering through a wild paradise of sandy dunes between the lagoon and the Mediterranean Sea. Today, it pains him to see the concrete-covered coastline crowded with jet skis—conditions that hardly seem to repulse the tens of thousands of younger tourists who flock there happily each year. Speaking more generally, the shifting baseline syndrome explains a lot of intergenerational conflicts as well: what was normal for our grandparents is no longer accepted as normal by their children or grandchildren, for better or for worse. Younger generations, for instance, are astounded to learn that their parents were allowed to smoke absolutely everywhere (doctors even did it in their offices), or that, if you wanted to hear from your sweetheart or faraway relatives, you'd have to wait several days for the mailman to bring you a letter. And so the baseline will keep shifting into the future: in a few decades, it will likely seem crazy to us that in 2017, nearly half of the 38 million automobiles on the road in France ran on diesel, creating a serious air pollution problem and leading to one of the biggest public health scandals of the early twenty-first century.

According to the German sociologist Dietmar Rost, who wrote a whole book about Pauly's concept and its tremendous ramifications, the main cause behind the shifting baseline syndrome is "a loss of historical depth."[10] To fix this problem, we must to learn how to stay in touch with the past while continuing to move forward. In the 1995 article he wrote for Bob May, Daniel dresses down fisheries biology for its collective amnesia, citing as a counterexample the science of astronomy, which "uses ancient observations (including Sumerian and Chinese records that are thousands of years old)." He goes on to urge his colleagues to take into account the observations of elders and ancestors, which until recently were relegated to the status of anecdotes. To illustrate this point, Daniel tells the story of Villy Christensen's father, a Danish fisherman who "reported being annoyed by the bluefin tuna that entangled themselves in the mackerel nets he was setting in the waters of Kattegat in the 1920s" when a market had not yet developed for that particular species. "This observation is as factual as a temperature record," Daniel argues, "and one that should be of relevance to those dealing with bluefin tuna, whose range now excludes much, if not all, of the North Sea."

Daniel wasn't the only one to have noticed a shift in baselines or to have suggested taking historical observations of marine resources into consideration, as anecdotal as they might seem. The idea was obviously circulating at the time, but he expressed and summarized it in a particularly eloquent fashion. His main source of inspiration was a

1984 book by Farley Mowat titled *Sea of Slaughter*.* Mowat, who passed away in 2014 a few days short of his ninety-third birthday, was a naturalist with the soul of a poet. At nineteen, he left his native Canada to fight in World War II and returned home from the Italian campaign with a case of chronic post-traumatic stress disorder that made him strongly averse to the violence perpetrated by mankind. Perhaps best known for his book *Never Cry Wolf*—adapted into a film with the same title—and his other writings about the Great White North, he is the author of forty books, mainly devoted to environmental causes and nature in Canada. An idol for environmental activists (Sea Shepherd named one of their ships after him) and ridiculed by academics who say that his lyrical prose lacks objectivity, Mowat, in *Sea of Slaughter*, provides a detailed and thoroughly researched analysis of the destruction of Eastern Canada's marine resources over the last five centuries.

Millions upon millions of birds, terrestrial and marine mammals, and billions of fish flourished in that lush coastal region before the arrival of Europeans. Going species by species, Mowat tells the story of a massacre of epic proportions, using historical texts and oral accounts from a vast network of informants spread from Cape Cod to the Labrador coast. The book reads like a thriller and leaves behind an unforgettable image—that of a bloodbath whose victims included seals, walruses, whales, polar bears, cod, and marine birds (such as the great auk, the northern equivalent of Antarctic penguins; unable to fly to safety, they were slaughtered by the tens of thousands and eventually went extinct in the mid-nineteenth century).

Thanks to Farley Mowat, we can imagine the incredible richness of these waters in the past—waters so dense with life that you could almost cross the ocean by jumping from the back of a large cod to a baleen whale, and so on, all the way to the other side. His writing also helps us understand the relative poverty of those same waters today, where only a small portion of the formerly abundant and diverse species of large vertebrates remain. In the late 1980s, Canadian fisheries

---

* Farley Mowat, *Sea of Slaughter* (Toronto: McClelland and Stewart, 1984).

scientists jeered at Farley Mowat's work, calling his historical analysis a fantasy. Those same experts, though, were also forced to announce the permanent closure of the Newfoundland cod fishery on July 2, 1992. That very fish had changed the course of history,[11] its stocks considered for centuries to be inexhaustible. The sudden shutdown of this fishery, which, with better management, could have gone on functioning indefinitely, put thirty thousand people—12 percent of Newfoundland and Labrador's population—out of work overnight. A quarter century later, the cod stocks still haven't recovered sufficiently to allow the commercial fishery to be reopened.

Following Farley Mowat's example, a number of ecologists have become interested in the history of fishing activities. Their goal, as Daniel explains it, is "to evaluate the true social and ecological costs of fisheries."[12] Protagonists abound. Though my choices are almost certainly biased, I will only mention two of them here. The first is Jeremy Jackson, a researcher who grew up on the East Coast of the United States, where he graduated from Yale in 1971. A towering, ponytail-wearing, earring-sporting character, Jackson initially became a world-class specialist in the evolutionary biology of coral reef dwellers. His work centered on Jamaica: "It was really a lot of fun for about ten years," he told the audience of his 2010 TED talk.[13] "The coral reefs [in Jamaica] were really among the most extraordinary, structurally, that I ever saw in my life . . . Then, in 1980, there was a hurricane, Hurricane Allen." The reef was destroyed, but Jackson and his colleagues published a paper in *Science* predicting it would come back quickly because "we know that hurricanes have always happened in the past." "And we got it all wrong," he adds tersely, "and the reason was because of overfishing." Indeed, the longtime absence of reef fish and the disappearance of sea urchins had led to the proliferation of seaweed and jellyfish, reducing both the coral's growth and the diversity of its forms. Jamaican coral reefs have, to this day, still not recovered from Hurricane Allen.

After that, Jackson immersed himself in the archives, studying the tragic history of the colonization of the Caribbean—and Jamaica specifically—in an attempt to imagine what those marine environments were like in the sixteenth and seventeenth centuries. He concluded that sea turtles, manatees, and monk seals must have been extraordinarily abundant and that they had a strong impact on the way Caribbean reefs and seagrass beds functioned. For example, there were an estimated 50 to 150 million green sea turtles living along the tropical coasts of the Americas before Europeans arrived. We can imagine the carnage that followed, not only to feed the sailors, but also the slaves they marooned on resource-poor islands. With an average individual mass of 450–650 pounds, these green sea turtles had a total biomass greater than that of the entire human population of the United States today.[14] Large marine animals have disappeared or dwindled into insignificance compared with the tens of thousands of individuals who lived before them. "Studying . . . reefs today is like trying to understand the ecology of the Serengeti by studying the termites and the locusts while ignoring the elephants and the wildebeest," Jackson concludes. Most significantly, he showed that the fish stocks available to the inhabitants of Jamaica were decimated during the mid-nineteenth century, whereas serious scientific studies of fisheries only began a century later. This obviously creates a shifting baseline problem, because "everyone, scientists included, believes that the way things were when they first saw them is natural."[15]

And so, starting in the 1990s, Jackson made the transition from evolutionary sciences specialist to historical ecology guru. Perhaps most importantly, he brought together nineteen American researchers at the National Center for Ecological Analysis and Synthesis in Santa Barbara with the goal of collecting all available information on the historical impact of fisheries. They used data from studies in paleontology, history, and ecology concerning both seaweed and coral-reef-covered coastal ecosystems and estuaries over periods ranging from a few decades to 125,000 years. Jackson and his colleagues concluded that "overfishing of large vertebrates and shellfish was the first major human disturbance to all coastal ecosystems examined" and "pollution, eutrophication, physical destruction of habitats, outbreaks of disease, invasions of introduced species, and human-induced climate change all come much later than overfishing in the standard sequence of historical events." The authors also noted that "the timing of ecological changes due to overfishing in the Americas and Pacific closely tracks European colonization and exploitation in most cases. However, aboriginal overfishing also had effects, as exemplified by the decline of sea otters (and possibly sea cows) in the northeast Pacific thousands of years ago." In closing, they remark that "the historical magnitudes of losses of large [marine] animals and oysters were so great as to seem unbelievable based on modern observations alone... The shifting baseline syndrome is thus even more insidious and ecologically widespread than is commonly realized."

Jackson and his colleagues' incendiary article made the cover of *Science* when it came out on July 27, 2001, causing a media frenzy by shattering two persistent illusions: one, the perception that the effects of overfishing were recent and limited to certain regions, and two, the myth of the pre-Columbian "ecological Indian" who lived in harmony with marine life in the Americas and Oceania.[16] Very politically incorrect at the time of its publication, the paper has become a classic of marine ecology, inspiring thousands of researchers around the globe.

One of the most eminent of these is Heike Lotze. I met Heike in 1993 at a pub on the German island of Sylt on the North Sea. We were both master's students at the time and we swapped ideas over some beers. Heike was in a funk: "I study seaweed—no one cares about that," she told me with a pout worthy of Jeanne Moreau. Primary research without any connection to the real world was *not* Heike's cup of tea, though, and she soon found a solution. Moving to Canada to escape the false meritocracy and latent sexism of German academic circles, she developed an interest in the historical ecology of marine environments and became a world-renowned specialist in her discipline. Today, she is a professor at Dalhousie University in Halifax. I knew Heike to be tireless and extremely meticulous in her work, so I kept close tabs on her career. She began with the Bay of Fundy in Canada, a body of water known for having the highest tidal range on the planet. For her thesis on the area, she assembled two hundred years of archeological and historical data, painstaking work that would mostly confirm the picture already drawn by Farley Mowat—and give it more "academic" substance. She also showed that overexploitation of marine resources, habitat destruction (caused by hydroelectric dams, coastal construction, dredging, et cetera), and various sorts of pollution come together to create a deadly cocktail for marine life. She reached the conclusion that her study area in the Bay of Fundy, where life had been so abundant in the past, "shows the most common signs of degradation found in highly impacted coastal areas worldwide."[17]

But what about the history of other coastal areas? Heike began by returning to Germany for a few years to do a synthesis on the Wadden Sea. This area, lauded by Erskine Childers in his novel *The Riddle of the Sands*, stretches from the Netherlands to Denmark and has a total surface area about half the size of Massachusetts. Made up of a labyrinth of islands as well as sand and mud flats constantly pounded by the tides, it is one of Europe's last wild regions, a harsh coastal zone on the list of UNESCO World Heritage Sites. Using her now-proven method, Heike assembled data covering the last two thousand years, exploiting neighboring countries' archives, archeological reports, and various scientific publications. Her conclusions were as follows:

> Humans have interacted with the Wadden Sea since its origin 7,500 years ago. However, exploitation, habitat alteration and pollution have strongly increased since the Middle Ages, affecting abundance and distribution of many marine mammals, birds, fish, invertebrates and plants. Large whales and some large birds disappeared more than 500 years ago. Although still of high natural value and global importance, the Wadden Sea is a fundamentally changed ecosystem.[18]

Indeed, who remembers the fact that gray and right whales were still abundant in those waters during the Middle Ages, as they were along many European coastlines?

Back in North America, Heike joined Jeremy Jackson's work group in Santa Barbara with the goal of coordinating a synthesis on the history of seven estuaries in the United States (including the Bay of San Francisco), two Canadian estuaries (including the Saint Lawrence), and three European seas (the western Baltic, the Wadden Sea, and the northern Adriatic) over periods ranging from several hundred to several thousand years. The sources they used were once again extremely varied, including lists of species whose bones had been found by archeologists in ancient rubbish heaps, sailors' accounts, ships' logs from whalers and fishermen, old nautical charts, even ancient art and old photographs, but also more recent studies of population size and genetic characteristics. Based on every area studied, the conclusion was overwhelmingly unanimous: "Exploitation stands out as the causative agent for 95% of species depletions and 96% of extinctions." And more specifically, "most mammals, birds, and reptiles were depleted by 1900 and declined further by 1950." When it comes to recent conservation efforts, Heike and her brilliant colleagues offer only lukewarm assurances: "Conservation efforts in the 20th century led to partial recovery of 12% and substantial recovery of 2% of the species, especially among pinnipeds, otters, birds, crocodiles, and alligators. Large whales, sirenians,* and sea turtles, however, remain at low population levels."[19]

Soon after, Heike and her partner Boris Worm wrote an even more ambitious report based on every historical study they could dig up on anything over a meter long that swims in the ocean: fish, whales, seals, sea turtles—a whole subset of species referred to as "marine megafauna." These 256 studies show that exploited populations have lost 89 percent of their historical numbers on average.[20] For instance, interviews with fishermen aged fifteen to fifty-five in the Gulf of California showed that older interviewees could name five times more species of fish and four times more areas where fish used to be abundant, and which are now overfished.[21] Over a longer period of time, archeological

---

* Manatees and dugongs.

excavations show that overexploitation by the Maori caused the distribution of southern fur seals in New Zealand to be reduced by 90 percent prior to the arrival of Europeans.[22] Here was a huge pool of data that not only confirmed the politically incorrect hypothesis of Jeremy Jackson and his colleagues, but also fought against the shifting baselines that were hiding the truly dilapidated state of marine ecosystems.

FISHING DOWN
MARINE FOOD WEBS

"I N THE LATE nineties, people were realizing that fishing is actually a problem for the oceans," says Daniel, whose face lights up as soon as we broach the subject of global fisheries. "To figure out if this activity is in the process of wiping itself out, you have to go beyond the Bay of Whatchamacallit or the Gulf of Whatever. When astrophysicists can't see something well enough, they build a bigger telescope. So that's what I did, build a bigger machine, the biggest one I could imagine—the world since 1950." With Ecopath and FishBase at his disposal, Daniel now had all the tools he needed to study fisheries on a planetary scale over a period of several decades.

But he also needed lots of data to feed those tools, which ran on information laboriously gleaned from the scientific literature, and his resources in Vancouver were limited. Daniel applied for funding from the Canadian government several times without much success: "I stopped asking them the day I received a response that said my application was excellent, but they still couldn't give me any funding." Initially, therefore, he had to rely on his students at UBC and his colleagues at ICLARM.

After studying the problem of trophic levels during the development of Ecopath and then again for his first article in *Nature* with Villy Christensen, Daniel realized that he could calculate the mean trophic level of all the fish caught in a particular zone over a certain period of time. For example, if a fishery mostly caught tuna, swordfish, and other superpredators, the mean trophic level would be higher than if it mainly targeted small vegetarian fish like anchoveta. As an exercise, he asked his master's student Johanne Dalsgaard to run such a calculation for the catches of all the fisheries in the northeastern Atlantic from 1950 to 1994. Johanne based her work on the FAO's fisheries statistics, which track catches according to species group, and the trophic levels of each of these groups, calculated using a series of Ecopath models. She found that the mean trophic level declined over time, "indicating that European fisheries catches are increasingly composed of fishes from the lower part of marine food webs."[1]

The results didn't come as a big surprise to Daniel, who knew that despite the best efforts of the International Council for the Exploration of the Sea, the northeastern Atlantic was still overfished, causing larger species to decline and forcing fisheries to fall back on smaller fish. These smaller targets initially multiply because the fisheries have eliminated their natural predators, but their stocks are soon over-exploited in turn.

Daniel had nearly forgotten about this analysis when, several weeks later, he found himself back at the Singapura restaurant in Manila with Villy Christensen and Rainer Froese. "I was having (low-trophic-level) Sambal squids with rice. What happened is that I suddenly realized that the plot by Johanne Dalsgaard could be easily expanded to cover the whole world."[2] Villy and Rainer were up for the challenge, and the team got to work with the help of Francisco Torres, who assigned a trophic level to each of the 1,200 species cited by the FAO, a task he politely characterized as "tedious." A few turns of the data mill later, they had their verdict: the trophic level of organisms fished in both salt and fresh water had declined since 1950 all over the world.

The paper where they presented their analysis "literally wrote itself," according to Daniel. He sent it to *Science*, the ultra-selective American journal. Their four-page contribution received a positive evaluation and appeared in 1998 with the title "Fishing Down Marine Food Webs."[3] Pauly and his colleagues emphasize that, despite this general trend, they also detected important regional variations, the drop in mean trophic level over time being stronger in the northern hemisphere where the fisheries are more highly "developed." They go on to underline the fact that "present exploitation patterns are unsustainable" and bring their article to a close with the following recommendation: "in the next decades fisheries management will have to emphasize the rebuilding of fish populations... within large 'no-take' marine protected areas."[4]

The article appeared in *Science* with the headline: "What is the future for fisheries worldwide? Bleak, according to an analysis by Pauly et al." An editorial accompanied the paper relaying the reactions of fisheries specialists, one of whom called it "a wake-up call." The text ends with a short remark from Daniel, one of those zingers for which journalists would come to love him: "If things go unchecked, we might end up with a marine junkyard dominated by plankton." Daniel, who already had a solid reputation in his field, became a celebrity thanks to that article, which is still the one he is best known for today.

I discovered the article at the library of the oceanographic institute in Kiel in February of 1998. "This guy is good," I thought to myself. "He's used open-source data from the FAO to do analyses that are simple yet revolutionary, and he's shown the world that everything marine ecologists have feared to be true for a long time is really happening."[5] His conclusions reminded me of a book I had read recently in which— already, in 1984—Timothy Parsons and his colleagues had announced that "no form of marine pollution is in any way comparable to the ecological impact which occurs with the removal of circa seventy million metric tons per year of predatory fish from the ocean ecosystem."[6] These words impressed me so much that I copied them down and pinned them to my dorm room wall.

The paper in *Science* was picked up by dozens of news outlets around the world but didn't make much of a splash in Daniel's native France. Still, he could hardly suppress his pride when the *New York Times*, which he read religiously, published a long article on his work in the Science section.[7] The *Times* journalist changed his name to David Pauly, but aside from that, Daniel found the piece to be "accurate, informative, and well written." In particular, the article explains that, according to Daniel, marine food webs will be "collapsing in on themselves" within thirty or forty years if nothing is done. The term "collapse," referring to both social and ecological contexts, appeared in that article years before the publication of Jared Diamond's book bearing that title.* To illustrate his point, Daniel tells the old story of the Newfoundland cod, pointing out that after that fishery was closed, the bottom scrapers known as trawlers shifted their attention to the cod's prey, shrimp. Once this new resource has been plundered, will fishers go after the shrimps' prey? Hard to do, since they mostly eat detritus—"It's mud," quips Daniel, concluding, "That's when you hit the wall beyond which the fishery has no more commercial value."[8]

In the media storm following the *Science* article, Daniel understood that "there are professionals with communication skills that most scientists lack." That was when he began answering calls from journalists: "I learned to communicate with them, so they could translate my findings into something that could be read and appreciated by wider audiences." One of his interlocutors from that period was Nancy Baron. An energetic Canadian journalist specializing in environmental topics, she interviewed him for a local Vancouver paper, the *Georgia Straight*, and later wrote one of the best portraits of Pauly: "Fisheries ministers, for the most part, do not read peer-reviewed journals, and so if scientists are to share their wisdom with the wider world, they have to come out of their ivory tower... Pauly is often the front man, venturing out on the tightrope across the abyss of the unknown... He is a storyteller... a master of metaphor."[9] Baron goes on to cite a series of Paulyesque remarks, including: "The [fishing] industry has acted like a terrible tenant who trashes their rental." And: "As a society, we don't have to do everything we can do. We don't allow people to drive over the speed limit to get somewhere faster, we don't allow machine guns to hunt deer, and we do not allow destruction of our

* Jared Diamond, *Collapse: How Societies Choose to Fail or Succeed* (New York: Viking, 2005).

marine resources." Nancy Baron sums up the impact of the article that Pauly and his colleagues published in *Science*: "While this paper is considered a classic amongst marine ecologists and fisheries scientists, what is more remarkable is that the concept has become widely cited by journalists and policymakers. The reason is that Pauly summed up years of work and mountains of data into a simple graphic that is now used far and wide, and a few oft-repeated catch phrases. 'Fishing down marine food webs' has become part of the vernacular in some circles."

But where is this magnificent graphic that Nancy is talking about? Don't look for it in the 1998 article—it only appeared later, as a drawing showing a cross-section of the ocean and its inhabitants from the surface to the sea floor. On the left, you can see this community as it should be, with superpredators like big sharks and tuna at the top; then, just below them, the smaller fish that they feed on; then the plankton; and finally, seaweed, starfish, and other creatures who dwell on the sea floor. As your eyes move to the right, you can see the community dwindle from the top down as a result of fishing—the sharks go first, then the medium-sized fish, and on the far right, we can see an overfished ecosystem with only a few small fish and plankton left. A diagonal arrow drawn from the top left toward the bottom right highlights the phenomenon of fishing down marine food webs. Several versions of this graphic can be found on the internet—an indication of its success—though they don't all give credit to the original creator.

Her name is Rachel Atanacio, but in Manila, everyone calls her Aque. I made her acquaintance at FishBase headquarters during my visit to the Philippines, and she blushed as I went on about how happy I was to meet the creator of such a famous illustration. Aque is a self-taught painter who has been drawing for as long as she can remember, with a marked preference for watercolors, landscapes, and portraits (her take on John Lennon is particularly impressive). Feeling she wasn't "made for an office job working nine to five," she started out as a volunteer drawing for different Buddhist reviews, which left her plenty of time to pursue her second passion, the violin. At a friend's suggestion, though, she joined ICLARM in 1988, then moved to FishBase, where she still works today. She is the one who drew, pixel by pixel, each species of fish that appeared in the first version of that famous ichthyological encyclopedia. Illustrating fishing down marine food webs was her first time working with Daniel. "I'm an artist, not a scientist," she tells me. "He had to explain to me exactly what he wanted, but I suggested adding the diagonal arrow that highlights the general decline." Aque used illustrated books about fish classification as references to make her drawings of the different fish species, then she taught herself how to use a graphic design program to copy the drawings and place them in the illustration. Daniel was more than satisfied with the result, and Aque has gone on to illustrate quite a few of his books and papers.

"Now that he's in Canada, Daniel has a lot of artists that can work for him, but if he's in a hurry, he'll send me an email and I'll get him what he needs the same day," she tells me with a smile. When Daniel makes his nearly annual trip to the Philippines to visit his FishBase colleagues, Aque and Daniel have a little routine: she drags him out of his office in the middle of the afternoon for a pensive walk around the rice paddies. In typical fashion, Daniel kicks off his shoes as soon as possible. "One day I said to him, 'You have very nice feet, smooth and slim,'" she tells me. "You know," Daniel answered her, "if I die, I want them to put me in a coffin and just show my feet sticking out, so that everyone can say, 'he's got nice feet!'"

His collaboration with Aque helped popularize Daniel's new concept, which he promoted with his usual sly humor. Nancy Baron remembers Daniel explaining to journalists that "having systematically stripped the oceans of the top predators, we are now eating bait, and we're headed for jellyfish . . . my children will tell their children, 'Eat your jellyfish soup.'"[10] Indeed, when fisheries remove the small, plankton-hungry pelagic fish, they are often replaced by jellyfish, which move in permanently because they eat the fishes' eggs and larvae, preventing their return. Daniel had seen quite a few studies from the North Sea, Namibia, and many other regions confirming the dangerous "rise of slime" in the oceans. This phenomenon leads to well-known problems for tourism and other industries, and is only made worse by climate change. So, Aque made a second version of her famous drawing, adding the jellyfish this time—and Daniel Pauly's provocative words led to a second torrent of articles in the press about the approaching "attack of the blobs" in marine ecosystems.

The fishing lobby was not amused by all this commotion, but Daniel knew that nothing could stop him from coming after them now. From a simple researcher in an organization linked, for better or for worse, to the fishing industry, he had risen to the mostly untouchable status of tenured professor. The University of British Columbia, under the enlightened direction of Martha Piper, would support him whatever happened. He therefore became increasingly outspoken in his critiques, even forging relationships with environmentalist NGOs. In Vancouver, Daniel was well placed to do just that—environmental activists have been organizing on the West Coast of Canada and the United States since the 1970s. Indeed, Greenpeace was born in Vancouver in 1971, where it launched its first expedition, to protest nuclear testing in Alaska. Among the founders of the association was Paul Watson, who would later launch Sea Shepherd in 1977.

To explain his change of tactics, Daniel pointed out the following: "When you study fishing, you notice that scientists, even those that work for a government, aren't often listened to. Environmentalist NGOs are the only ones who can take scientists' conclusions, explain them, and force politicians to take them into account. Those in power make reasonable decisions when they are forced to do so." When it comes to NGOs, he admits that "with them, it's another dance, because they want us to become advocates, which I can't do because I'm a scientist. So, I keep publishing, but I do it so that nonprofits can use my work. I think I've managed to do just that because they cite me a lot. It's a dangerous liaison, but it works. On the other hand, very few scientists act so openly. Most of them are afraid of NGOs."[11]

Such fears are often justified. Daniel, who has always been subject to hostile reactions from his colleagues, especially those he has criticized mercilessly in the past, was increasingly the target of scientific and political attacks. "After the article in *Science*," Francisco Torres tells me with a nervous laugh, "we got praised here, criticized there, bashed here, appreciated there." The many letters of protest and a few threats emanating from the northeast coast of the United States aside, the first major attack on Daniel and his ICLARM colleagues' work came in the form of a response published in *Science*. Such responses in *Science* and *Nature* are a significant feature of scientific publishing, and they appear for one of three reasons, which I will mention in order of importance. First, jealousy: nothing could infuriate other fisheries scientists more than to see Daniel—the unruly problem child—land a revolutionary paper in the best scientific journal on the planet and then benefit from the media attention to become a new alter-globalist icon. "The FAO was mad as hell," comments Francisco Torres. Second, strategy: by sending a response to *Science*, the other side often hopes to score a publication in the same journal, "winning" a spot in a prestigious journal with little effort. And finally, integrity: even if they are in the minority (at least, in my opinion), some scientists are honestly interested in academic debate and write to *Science* because they have detected evident bias in a recently published study.

"By far the most interesting critique of 'Fishing Down Marine Food Webs,'" Daniel notes, "was by a group of FAO staff who suggested that the FAO catch statistics used for our contribution were not detailed and reliable enough to support the inferences drawn from them. (I don't dare to think what they would have written, had we written such a thing about the FAO statistics!)"[12] The FAO scientists' commentary begins with praise for Pauly and his colleagues, saying they "are to be congratulated for giving this important issue high profile." However, they go on to insist that the FAO does not agree "that a general decline in mean trophic level of marine landings is likely to have occurred in many regions." The authors express several objections that they say seriously limit the study's validity. Daniel, Rainer, and Villy responded point by point, and the two letters appeared side by side in *Science* in 1998.[13]

What follows is a brief summary of their game of academic ping-pong. First objection: the trophic level of fish caught is unknown for

70 percent of saltwater species and 40 percent of freshwater ones. Additionally, it can change according to the age of the fish and the season. Response: even if the FAO's dataset is far from perfect, it was in the zones for which the most detailed information was available that our analysis showed the sharpest decrease in mean trophic level. Second objection: FAO statistics cover the volume of the catch, not the biomass of the fish of different trophic levels present in marine ecosystems. Response: this isn't a problem because fisheries are now so global, exploiting stocks in every corner of Earth's waters, that the relationship between volume fished and volume present is a robust one. Third objection: when it comes to freshwater fisheries, the FAO statistics don't differentiate between the volumes of wild-caught fish and those produced in aquaculture, despite the fact that this activity is growing quickly and has an increasingly large impact on mean trophic levels. Response: we reanalyzed the data without the freshwater fisheries, and our conclusions are the same. Fourth objection: the eutrophication of coastal waters—that is, their enrichment by dissolved nutrients from agricultural runoff—boosts plankton production and causes a population explosion among organisms with low trophic levels which has nothing to do with fishing. Response: the fishing-down effect still appears when species with low trophic levels, likely to become more abundant in coastal waters, are omitted from the computation of mean trophic levels.

Though his counterarguments were solid, Daniel did not take the scientists' criticism lightly. On the contrary, it steered his research for years to come as he worked to find additional evidence of planetwide overfishing: "As it turned out, each of Caddy et al.'s points reinforced our original conclusions. In effect, they became 'judo arguments' (Asimov 1977), whose detailed examination strengthened the case that we were 'fishing down the food web.'"[14]

Pauly and his colleagues end their defense on a conciliatory note, emphasizing the FAO's efforts to document, and try to halt, "the excessive global fishing capacity that has depleted major fisheries" as well as "the vast effort that went into generating and maintaining the global FAO database of landings." Pauly's team seemed to have won this round fair and square: in the decade that followed, the concept of fishing down marine food webs only gained more traction, its validity confirmed by twenty-eight studies performed off the coasts of five continents and on a global scale. Recognition also came from an unexpected source. In 2004, the Convention on Biological Diversity, meeting in Kuala Lumpur, identified a number of biodiversity indicators that would allow them to track progress toward significantly reducing biodiversity loss by 2020. Among the eight indicators to be implemented immediately was a "marine trophic index" very much inspired by Pauly's team's work. "Of course, we were very proud of the fact that our findings had ultimately led to an important new policy initiative."[15]

But scientific debates are never really over, and, in 2010, the one on fishing down marine food webs burst back onto the scene. Trevor Branch of the University of Washington in Seattle and eight other scientists published an article dryly titled "The Trophic Fingerprint of Marine Fisheries" in *Nature*—a very precise attack designed to shatter Daniel's conclusions on the topic.[16] Using a combination of mathematical models and what its authors call "global assessments of mean trophic level" based on catch volumes and estimates of the size of fish stocks, they reach the conclusion that the trophic level of catches does not reflect structural changes in marine ecosystems. The authors recommend abandoning this particular indicator, but their ideas for how to replace it are somewhat vague. For them, the priority should be "measuring and reporting changes in marine biodiversity by tracking trends in abundance relative to reference points for conservation and sustainable use." They suggest developing new methods to that end that could be used by "countries with few resources for science and assessment." They could have added, "And good luck with that!"

When he saw Branch's paper in *Nature*, steam came out of Daniel's ears: "I was shocked, just stunned by their nerve," he recalls. In fact, Trevor Branch had shown Daniel his preliminary results several months earlier. "Products," Daniel says, "with trophic levels that rose but then went back down again over time without showing any general trend." Daniel alerted Branch to the fact that his data did not differentiate between coastal and offshore areas: in fact, because coastal zones were overfished first, the mean trophic level of catches there dropped. Later, though, fisheries shifted to offshore zones that were still rich in fish, causing a brief jump in mean trophic level that only dropped off again over time as resources farther from the coast were depleted in turn. "As far as I was concerned, that conversation was the end of it," recounts Daniel. But, obviously, that wasn't the case.

Branch's surprise attack also relied heavily on data from North American and European fisheries, adding only a few African, Asian, and Australian examples to make the analysis appear more "global." This approach, centered on the Global North, left Daniel seriously exasperated. "You can fit sixteen Alaskas into the African continent, so the idea that Alaska can serve as an example for stock management is totally absurd—Africa should be the reference point," insists Daniel. But the straw that broke the camel's back was Reg Watson. A highly skilled analyst, Reg coauthored Branch's article despite the fact that he had been a close collaborator of Daniel's for the last ten years at UBC, where they worked together on the same project—a project from which Reg supplied part of the data for Branch's analysis. His defection came without warning.

"I almost panicked at first," Daniel remembers. "Of course, I looked like an idiot—like they'd caught me red-handed. I didn't know what to say because anything I did say would just be angry words. Hasty replies are like pissing into the wind; they just come back and hit you in the face. To regain the upper hand, we had to do studies showing that their claims were inherently false." Daniel formed a new team with his postdoctoral student, Kristin Kleisner, and her husband, Hassan Mansour, "a guy who was good with numbers." Together, they developed a model that took into account the spatial expansion of fisheries over time and the resulting impact on the mean trophic level of the catch. "If it isn't done that way, it would be like a farmer saying that he had increased his farm's productivity when in fact he just increased acreage," Daniel points out.

But that argument wasn't enough to convince Trevor Branch, who instead proposed a diplomatic solution that smelled of half-hearted consensus building.[17] In his mind, there were four working models for how fisheries develop over time: fishing down marine food webs (as described by Pauly and his colleagues), "fishing up marine food webs" (as if fishers went after mussels, then fish, then whales), "fishing through marine food webs" (where all trophic levels are exploited simultaneously), and finally "fishing for profits" (where marine organisms are targeted according to their commercial value, regardless of their trophic level, as is the case with abalone, for example).

FISHING, FISHING, AND more fishing—Daniel obsessed over his work and the verbal and written skirmishes that came with it, so much so that he wasn't always around to see his kids grow up. "I have regrets," Daniel admits when I ask him about his children, an even thornier topic than his academic battles.

"I don't envy you the task of writing about them," Rainer Froese tells me. "Depending on what comes out of it, Daniel might pin you to the wall." Message received loud and clear. Obviously, I know my interviews with Angela and Ilya will be among the most sensitive of my entire investigation. I begin prudently with the least tricky of the two interviews and go in search of Angela Pauly. Her parents give me an address in Brussels, and I arrive in that European capital not long after the terrorist attacks of March 22, 2016. "That day," Daniel tells me, "I was traveling in the middle of nowhere in the Yukon and we didn't hear from Angela for several hours. Sandra and I were very scared."

Finding a hotel room proves easier than usual and a strange stillness reigns over the city. I sit on a café terrace, sipping a beer and chatting with a Belgo-Moroccan couple who apparently haven't lost their sense of humor. "We were going to take a trip abroad," they tell me, "but then we said, what's the point when we can get blown up at home!" Angela Pauly, on the other hand, is not one to joke about terrorism. The European powder keg has her worried, but she likes Brussels. An elegant thirtysomething, she lives on the top floor of a historic building and works for an international marketing and communications agency. Her online profile says she's "ridiculously positive." Angela answers my questions calmly, with just a hint of humor. Like the rest of the Pauly family, she gets straight to the point: "When I was little, I wouldn't recognize him when he'd come home from a trip, especially if he'd shaved his beard. My father is an academic; his work defines him. I know it has nothing to do with what he thinks of us, even if I probably felt differently at thirteen... My childhood was happy, but there was a lot of pressure to succeed in school. It was subtle, but it was there. I was a good student, but not someone who wanted to work all the time. I was happy with an A; I didn't need an A+. My parents were frustrated that I didn't push myself harder." I am led to understand that this academic elitism didn't just come from Daniel, whose "life was saved by education," but also from Sandra, a child of the fight for civil rights whose family was no less than extraordinary. "The Wade sisters are highly intelligent," Angela says, "and their mother always wanted them to be independent."

Angela grew up in an upper-class neighborhood in Manila, mostly sheltered from the political turmoil of the Philippines. "There's a huge difference between my father's childhood and my childhood, I know that. But even in the Philippines, I knew I was different—people stared at me, they touched my hair. Luckily, I never had any bad experiences and it didn't affect the rest of my life. I've been very lucky; I have a lot of friends." As a child, Angela had everything that Daniel was denied in La Chaux-de-Fonds and more: loving and enlightened parents who spoke to her in two languages, the apartment with the terrace and the pool, and the comfortable life of a well-to-do expatriate family with horseback-riding lessons and weekends at the beach, school vacations in California and France, and an excellent education at the international school where Sandra worked. "After her baccalaureate in 1999," Daniel remembers, "Angela was accepted into all the big American universities, and she chose the only one that didn't offer her a scholarship, where the annual tuition was over thirty thousand dollars." Angela adds, "I didn't know it was possible to not go to college—I didn't think twice about it. And because I didn't grow up in the United States, at first I didn't know anything about African American culture."

Angela went off to Wellesley College, one of the most prestigious women's colleges in the United States, ranked number four in the country for social sciences. There, Angela became part of what is often called "the most powerful network of women in the world." But the experience wasn't all positive: "It's an amazing school, but I was very unhappy my first year there because of the culture shock. I had an American accent, but I was *not* American." Angela worked while she was in school to cover part of her tuition and spent more time at her boyfriend's place near MIT than she did on the Wellesley campus. After four years in Boston, Angela followed her boyfriend to San Diego, where he joined the Navy. "For a year, I didn't do much. I got a job selling shoes, I went to the beach. My father was not happy." One day, she got a call from Daniel: "I didn't pay for Wellesley so you could work at a shoe store—you're going to go to graduate school in Vancouver." Angela did as she was told: "I'm a good daughter. I do what my parents ask me to do."

Sandra Wade-Pauly, meanwhile, with no reason to stay in the Philippines now that her children had flown the nest, found a job in Vancouver and joined Daniel there. As a result, Angela found herself living with both her parents during the year and a half it took to complete her master's at UBC. Daniel sighs happily as he describes that "fantastic time" during which he spent a couple of hours each day with his adult daughter. Then she left again, off to launch a career that, like her parents', would take her across continents, to Washington, DC, Costa Rica, Napa Valley, Madrid, and finally, Brussels. Daniel's chest swelled with pride when Angela joined Oceana, an American NGO dedicated to ocean conservation, and he frowned with disappointment when she left to work for a multinational corporation. "Gangsters, that's what I call them," he says. "Those people who do public relations for dictators!"

"At first," Angela admits, "I didn't like the idea of working outside of the nonprofit sector. We do have clients that I'm not very happy about. Morals are flexible—we don't work for the cigarette companies, but we do work for the Turkish government." She adds, "I was very anti-corporate, but I'm not anymore. My father is much more dogmatic about it than I am. I'm not going to go to work for Shell or Total, but if I have a car, I'll put gas in it even if it makes me uncomfortable."

I bring the conversation back around to Daniel Pauly's work on overfishing. "I think he's right," Angela tells me. "And I'm not just saying that because he's my father. I've worked in that field and I've read the literature. He's right about it, and it's good that he's gotten some recognition because it has to be very frustrating for him. Overfishing has to stop, not because fish have feelings, but so that we can feed everybody. I wouldn't say that my dad is a genius, but it's crazy how much he knows, how much he's read, worked. I admire what he does, but, seeing him, I know that it's not what I want to do. It's not as important to me."

When I ask about her relationship with her father now that they live so far apart, she answers quickly: "When we talk on the phone, he asks me about work—that's his way of letting me know that he worries about me. Incidentally, in 2013, when I left my boyfriend at the time, my dad called me at the office to ask how I was doing. It wasn't like him, but it was really sweet. Usually, he's the last person I would talk to about that sort of thing, but he made an effort, he gave me a call."

When it's time for Angela to return to work, I walk her back to an official-looking building, which she enters with all the confidence of young upper management, her high heels breezing effortlessly around a corner and out of sight. Before we said goodbye, though, she made one final remark: "You know, even if my father wasn't around a lot, in the end, the most important thing was that I felt loved. Sounds cheesy, but it's true."

AS DANIEL PUTS it, "You've met my favorite daughter, and now you're going to meet my favorite son." But Ilya lives in Montreal and I won't see him until almost six months later, on a cold November evening. I've already heard a lot about him, in Vancouver and in France, and it became clear that his path had not been as smooth as his sister's. "Ilya is a fantastic person," Angela says, summing up the situation, "but his teenage years were a little different."

I fly in from New York with a bad case of bronchitis and make my way to downtown Montreal, where I wait for Ilya to get off work. I am a little nervous, mostly because I've lost my voice and have to do a conference on my research the next day at McGill University. But I also wonder what right I have to dig into the tumultuous youth of Ilya Pauly, who probably has better things to do than chat about his father's illustrious career. Ilya's emails were somewhat terse, but he welcomes me politely, cooking dinner while we talk. His girlfriend, Kasia, a social worker of Polish heritage, smiles gently at him and serves me copious amounts of herbal tea. With his almond-shaped eyes and long hair, Ilya reminds me of a Native American. Just a bit taller than his father, he isn't quite as towering as he looked in the photos. Their building is quiet; jazz plays in the background, and Ilya's Quebecois accent has a musical quality to it—despite the frog in my throat, this interview turns out to be my most radiophonic.

"Starting in the early nineties, I was aware of my father's absence. Sometimes, he would be gone for months, and toward the end it got to be stressful. My mother took it badly and so did we. He definitely made a lot of sacrifices when it came to his family. He came back with gifts and stories; he was always good at explaining what he did." Ilya grew up in the same sheltered environment as his sister but didn't go to the same school. "I absolutely *had* to go to the French school, which I thought was terrible. As a ten-year-old, mixed-race kid living in the Philippines, I was supposed to be learning about how the Gauls had invented everything! Daniel probably taught me too much, too young for me to be able to swallow what they taught at school." When I'd spoken to Daniel about it a few weeks earlier, he'd sighed heavily: "Ilya refused to accept the rules, and I blame myself for encouraging him in the beginning. When he'd say 'this is what they taught me in school' and it wasn't right, I would tell him. I didn't realize that what I was doing was destroying the teachers' authority. Those young volunteers who worked at the French school were such idiots—they made negative comments about interracial marriage and presented bilingualism as a handicap. What they really resented was that their students spoke English better than they did." Things got so out of hand at school and at home that the Paulys sent their then adolescent son to a private school in Singapore. Ilya laughs when he tells me that story: "I climbed over the wall so often at that place that after the first two weeks, they gave me the keys, just so I wouldn't fall and break my neck. I messed around instead of going to class, like any fourteen- or fifteen-year-old kid would do." Listening to him, I wonder if Daniel had already told his son about his own unruliness at that age, back in La Chaux-de-Fonds. Daniel must have been thinking of his own history, though, because around that time, he offered to help Ilya find his biological mother in Indonesia. "But I already had too many things I needed to figure out in my life," Ilya tells me. "As far as I'm concerned, my mother has always been Sandra."

According to Daniel, "Ilya's way of refusing to do things went pretty far... As soon as he suspected you were trying to teach him something, it was 'no'—I'm not even sure he realized he was doing it." In the early nineties, their father-son relationship was a minefield, but there was always one way to call a truce: diving, maybe because Daniel couldn't scold his son underwater. He didn't have much to teach him about the sport, either—an accomplished swimmer by ten years old, Ilya was a seasoned free diver by the time he was a teenager, and an instructor at seventeen. "Back then, I could easily dive down twenty meters [sixty-six feet] and stay under for two or three minutes," he tells me matter-of-factly. After his brief stay in Singapore, Ilya dropped out of school and half-heartedly took some long-distance classes. Mostly, he took part in diving expeditions, exploring out-of-the-way corners of the Philippines. On a personal level, though, he was adrift. "I should have spent more time diving with my son," affirms Daniel—at the time, though, he mostly studied the Pacific from thirty thousand feet above the waves.

At his parents' request, Ilya joined Daniel in Canada. The atmosphere in British Columbia came as a shock. "The hardest period of my life," Ilya admits, "was when I lived in Vancouver. I would go days without speaking to anyone. The people are polite enough, but there's always distance—it's an English thing. In the Philippines, though, it's the exact opposite. You talk to someone for five minutes and he'll invite you to his house." The contrast was that much starker because Ilya had just spent six months on a tiny Pacific island, "literally one kilometer by five hundred meters,"* living in a tent and going diving six days a week with a group of twenty other people, "really laid-back guys, funny, like a little family." Ilya arrived in Vancouver in September and lasted eight months under the city's semi-constant cloud cover. "I was left to my own devices," observes Ilya. Daniel remembers it as an "infernal" time. His son was expelled from three of the four institutions where Daniel had enrolled him, and he spent his time wandering the city "in bad company."

"Any time I managed to have a real conversation with someone, they were from Montreal," Ilya remembers. So, at the age of twenty, he left his father and moved to Quebec. "I couldn't stop him," Daniel sighs.

"I did all kinds of odd jobs in Montreal, anything I could find, including washing dishes," Ilya laughs. "I went from diver to dishwasher!† At eighteen, nineteen years old, I thought I knew everything there was to know, but when I got to Montreal, I realized that wasn't the case," he admits. Ilya discovered a passion for music and studied at a school for sound engineers before becoming an instructor there himself and eventually opening his own studio and working as a DJ.

Today, he mostly makes his living working for an audiovisual equipment company. And twenty years after moving there, Ilya is still in love with Montreal: "Where I live now, I don't have to tell anyone where I'm from. Francophones can be a pain, but at least they're up-front with you." He's also made peace with his demons—and his father. "All the pressure I was under, it's something I regret in life, but mostly in his life.

---

* About 125 acres. (TN)
† Ilya is making a pun: in French, the words for "diver" and "dishwasher" are homographs. (TN)

He's also trying to wake people up, so I can forgive him for a lot, for being away so much, just because of that." Ilya is still sorry to say that his father "doesn't unplug very often—science defines him and he's afraid of anything unscientific. He's not at all esoteric, you could even say he's anti-spiritual!" Aside from that, Ilya says, Daniel "also doesn't pay much attention to social norms and he expects to be listened to, which isn't always appropriate with your family."

But Ilya also insists that his father has always been his "moral compass"; he squeezes him "like a lemon" when they see each other (usually over the winter holidays in California). "My father also taught me to think about how the world works. For example, he was the first person to talk to me about the war in Syria as a climate war." Their favorite topics of discussion sound like something straight off a Philosophy 101 syllabus: "Why is progress dead?" They also talk a lot about feminism. "I'm a feminist like my father," Ilya declares. "Part of him is totally allergic to masculine energy, to anything competitive. He's so anti-macho that I don't even think he knows what it is."

After a final cup of tea, we say our goodbyes and Ilya, who knows that my next trip will be to the Philippines, gives me some advice: "Don't smoke any joints while you're there! They've just elected a murderer who executes junkies—it's crazy over there right now." Coming back around to Daniel and my work on the biography, he observes, "My father spends so much time looking inward, I think that seeing himself reflected in something like this will be good for him."

THE SEA
AROUND US

RACHEL CARSON DIED of breast cancer at the age of fifty-six, and her life was not always an easy one. Fascinated by the natural world she discovered on her family's farm in Pennsylvania, obsessed with poetry and in love with her English teacher, Rachel became her family's breadwinner at an early age and was forced to abandon the career in ecology to which she had aspired. As a teacher, then an unassuming lab tech, she scraped enough money together to go to graduate school and earned her master's degree in 1932 from Johns Hopkins. Her thesis had focused on the embryonic development of fish, so she quite logically went to work for the US Bureau of Fisheries, where she was soon recognized for her literary talents. She began editing all the bureau's reports and writing a weekly radio program and newspaper articles designed to stoke the public's interest in aquatic life.

Carson was a prolific author—her style precise when explaining scientific information, but steeped in a lyricism that reflected her sense of wonder and her intellectual enthusiasm for the beauties of the natural world. Editors soon discovered her work and in the 1940s and '50s, she wrote a series on marine environments, including *The Sea Around Us*,* the book that would make her famous around the world and give her a degree of autonomy. She left her government job to write full-time and began enjoying a mostly platonic—and not entirely

225

---

* Rachel Carson, *The Sea Around Us* (Oxford: Oxford University Press, 1951).

secret—relationship with a married woman, Dorothy Freeman. The two sweethearts exchanged nearly nine hundred letters, an archive that provides valuable information about Rachel's literary activities throughout the 1950s. Obsessive like all great artists and intellectuals, she collected thousands of bibliographic sources and drove her editors insane with constant delays for years at a time. But it all paid off in 1962 when she finally published her most controversial work, a seminal text of the burgeoning environmentalist movement and a project she'd been hatching for a long time: a book called *Silent Spring*.*

Starting in the 1940s, she had edited the first reports about the impact of DDT on fish—but decision makers at the time didn't even bat an eye. Twenty years later, though, Carson sallied onto the public stage with a thorough summary of the research on the environmental consequences of DDT and other pesticides. The multinational corporation DuPont and other manufacturers of chemical death attacked her and her book, but *Silent Spring* was carefully researched and stood up to criticism. Soon, President Kennedy ordered an investigation. Carson also spoke publicly, including on television. She was ultimately vindicated and covered in honors, a shower of glory that would prove all too brief. Even as Rachel Carson wrote about the links between pesticides and cancer, she herself suffered from an "environmental cancer" that she knew would soon end her life. A few heart-wrenching letters to Dorothy later, she died alone on April 14, 1964. .

Daniel discovered *Silent Spring* in German translation in 1968 when he was thinking about becoming an agronomist and working in Africa. The following year, during his first semester at the University of Kiel, a higher-up from BASF, one of the largest chemical manufacturers in the world, came to do a presentation for the students. The *Herr Professor* called Rachel Carson a hysteric and explained coolly that DDT and other chemical products used in agriculture had never hurt a soul.

---

* Rachel Carson, *Silent Spring* (Boston: Houghton Mifflin, 1962). The book predicts a silent springtime throughout the country due to the disappearance of birds and insects that will all have been killed off by pesticides.

"According to him, we could have eaten them," Daniel jokes. "It was such a big lie that—along with the Nazi professors—it pushed me to change my major and study oceanography." Daniel never forgot Carson's courage and brilliance, however, and three decades later, she came back into his life at the dawning of one of his greatest endeavors.

VANCOUVER, FALL 1997. For several years now, Daniel had been extricating himself from the more institutional side of fisheries biology and getting closer to environmentalist organizations. This radical transition has allowed Daniel to speak more freely about his work—though some of his colleagues saw it as a betrayal. His display of ideological colors also earned him an invitation to one of the Pew Charitable Trusts' work groups in Philadelphia. The American NGO, named for its founder, businessman Joseph N. Pew, mainly funds cancer research and the American Red Cross, but had recently shifted its focus to environmental issues. Though Daniel wasn't yet a big name in his field, Joshua Reichert, head of Pew's environmental programs, managed to spot him. Reichert gave his panel of experts a single day to answer two questions: "Is it possible to assess the health of the oceans regularly, and, if yes, how?" and "Is it possible to determine which factors are contributing to the condition of the ocean and their consequences for marine life and for humanity?" As Reichert later wrote, "With one exception, the members of the group said that it was not doable without very large investments in monitoring technology, for which there was no recognizable donor other than governments, which would be unlikely to provide the funds needed. The one exception was Daniel Pauly, who, in his indomitable fashion, said that it was possible to undertake this kind of analysis for ocean fisheries."[1]

A mischievous grin plays across Daniel's face as he tells me the story of that memorable meeting. "I didn't have the same level of credibility as the other researchers they'd invited, and I was the youngest one there. So, I let them speak and I arranged it so I would be the last to give my presentation. I gave a summary of what had been said to show they'd all come to more or less the same conclusion, and then I suggested using the FAO's fisheries data, which is a huge and freely available source of information, to understand the impact of that economic activity on the world's marine ecosystems over the decades." That afternoon, Reichert saw off those highly distinguished scientists with perfect politeness but kept Daniel back for a quick discussion, one-on-one. Their intellectual compatibility was obvious right away. "Josh is all academic," Daniel tells me, describing a mustachioed man in a three-piece suit. "An anthropologist, studied at Princeton— he can come across as standoffish, but actually, he's a romantic." The gray-haired scholar asked Daniel to write a grant for his bold research proposal. "I confess, the dossier I submitted wasn't very clear," Daniel admits. "Reichert's whole team told him not to fund me. The seven American researchers he sent my work to said the same, except for one who thought it was a good idea but wanted funding for his own team to do it." Reichert had no one to back him up but enough power to follow his instincts and go through with the project. The trust gave Daniel and his team 1.2 million dollars per year on the condition that they begin by analyzing fisheries in the North Atlantic—the area for which the most information was available—and submit an exhaustive research report after two years.

Daniel was delighted, baptizing the program "Sea Around Us" in honor of Rachel Carson and her sweeping vision of underwater worlds and humans' impact on the environment. Launched in July 1999, the Sea Around Us began by punching up the UBC Fisheries Centre, "showering money" on five or six researchers close to Daniel, including Villy Christensen, who had recently followed his friend and colleague to UBC. Once they'd recruited an army of students and consultants, the party could begin. "At first, we didn't have a strong sense of direction," Daniel admits, "but Josh came to see us and gave us clearer objectives." The team doubled down, working to produce a vast portrait of the ecology of the North Atlantic over the course of the twentieth century.

For the region in question, the FAO had recorded the quantity of fish caught each year by species, but without any precise information on catch location. The tricky part would be reconstructing that missing data. Reg Watson performed the necessary analyses, using species distribution data from FishBase to create the first catch location maps to cover North Atlantic catches from the 1950s to the 1990s. They show a clear increase in pressure from fisheries in coastal waters over time, as well as gradual expansion into offshore areas and along new coastlines, particularly in West Africa.

Next, a battery of Ecopath models built under Villy Christensen's supervision made it possible to estimate the biomass of fish present in the environment for the whole North Atlantic over the course of the twentieth century. By calculating the relationship between mass fished and mass present, the researchers showed a troubling increase in fishing pressure over time and, in particular, overfishing in Western European and Scandinavian waters since 1975. They drove their conclusions home by using similar methods to demonstrate that populations of large fish had decreased by 90 percent in the North Atlantic during the twentieth century, and that more than two-thirds of this drop in numbers had happened after 1950.[2]

Joshua Reichert had ordered a health assessment of the North Atlantic, and Daniel and his colleagues gave it to him. "We had shown that the biomass had collapsed; it was easy to see that the Atlantic wasn't doing well," Daniel concludes. This time, everyone at Pew got behind their work, and the Sea Around Us's funding was renewed year after year. The results of their work on the North Atlantic initially appeared as a series of reports in 2001, then as scientific publications in the years that followed—for example, Villy Christensen et al.'s article on the decline of larger fish, published in March of 2003.

But the Sea Around Us researchers weren't the only ones working on overfishing. Less than two months later, in May 2003, a Halifax-based duo made waves in the media that would be felt around the world. These new protagonists were Ransom Myers and Boris Worm. Myers, the son of Mississippi cotton farmers, was a well-known figure in the small world of fisheries biology. He was tall, brown-haired, and brown-bearded; some people thought he looked like Thor, the god of thunder,[3] but my French colleagues called him "the Ayatollah." After wandering the Middle East and Africa during the 1970s, he crossed the Atlantic in a 27-foot sailboat to study mathematics in Canada. Hired by the Canadian Department of Fisheries and Oceans after his doctorate, he moved to Newfoundland, where he saw the tragic collapse of the cod stocks unfold in real time. As Jeffrey Hutchings, his friend and colleague at the time, would tell it, "You could not work at the Department of Fisheries and Oceans in Newfoundland in the early 1990s... and not be cognizant of the obligation that scientists have to society of communicating their research widely, publicly, and honestly."[4] Myers took a stand, telling the press that neither ravenous seals nor warming waters were responsible for decimating the cod stocks—obviously, overfishing was. The Department of Fisheries reprimanded him severely, but they couldn't shut Ransom Myers up. He turned his back on the ministry and became a professor at Dalhousie University in Halifax in 1997. It was there that, three years later, he received a visit from a young German researcher, yet another graduate of the oceanographic institute in Kiel.

Boris Worm is from the Black Forest, and he does indeed project something of the woodsman-intellectual, like a young Aldo Leopold.* When I first heard about him in the late 1990s, all the girls at the institute couldn't stop talking about his curly blond hair and his adventures in the Yukon and on Vancouver Island, where he spent his summers observing the orcas. Our professors had also fallen under the spell of this obviously gifted student, who was a hard worker and detail oriented. University legends abound when it comes to Boris Worm: one of them concerns his time at the Roscoff marine research station in Bretagne, where his whole class went to study marine organisms on the rocky French coast. Boris had taken an interest in the feeding habits of sea snails and, unsatisfied with the information in his textbooks, took to sitting on the ocean floor in his diving equipment for hours at a time, patiently waiting for the tiny creatures to give up their secrets.

---

* See: Aldo Leopold, *A Sand County Almanac* (Oxford: Oxford University Press, 1949).

Boris was not a fisheries biologist, however. "When I was a student, I always thought that subject was so boring," he tells me. "It just seemed like accounting to me." Indeed, he still loves the intertidal, that part of the coast where the tides go in and out and where he likes to test complex hypotheses about interactions between different species. But after his doctorate in Kiel, he began looking for a job and met Ransom Myers. "We hit it off and he took me to a big fisheries conference in Florida," recounts Boris, who admits that he was "very impressed by the sophisticated statistics in certain analyses." He moved to Halifax and the Worm-Myers duo soon began using those "sophisticated statistics" to study interactions between species in the parts of the North Atlantic where cod were fished.[5] They were able to demonstrate that once all the cod had been caught, shrimp became abundant because there were no more cod to eat them. The fisheries that had targeted cod therefore quickly shifted to shrimp, neatly demonstrating the concept of fishing down marine food webs articulated by Daniel Pauly.

Then, along with four other researchers at Dalhousie University, they used data from the American longline fishing fleet, which tracks accidental catches of sharks in the Gulf of Mexico. By studying the number of sharks captured year after year, they were able to show that their numbers had declined by 75 percent in fifteen years. The resulting paper made headlines in *Science*,[6] but Boris and Ransom were just getting started. In fact, Ransom had spent the preceding decade gathering information on the abundance of large fish (cod, tuna, swordfish, et cetera) for four coastal regions and all of the temperate, subtropical, and tropical zones in three oceans. This data was mainly mined from the records of Japanese long-liners, which track the precise number of fish captured by species per year, per region, as a function of the number of hooks put into the water on longlines that can reach up to 62 miles in length. This allowed Ransom and Boris to reconstruct how the biomass of groups of exploited species had changed over time by using the number of individuals captured per one hundred hooks as an indicator.

"Long-lining has expanded globally... like a hole burning through paper," Daniel quipped, coining a phrase that would be picked up by

media outlets the world over.* Indeed, the maps that Boris pulled up
on his computer screen for the television cameras demonstrate that
although in the early 1950s the Japanese fleet kept to the Pacific, where
they made substantial catches north of Australia, by the end of the
decade, they had moved out into the warm waters of the Indian and
Atlantic Oceans, carrying out huge raids in the waters off Brazil and
West Africa. By the 1960s, their conquest of the waters inhabited by
tuna and similar fish was complete—and catch rates in the Pacific were
already falling. This trend was confirmed in the 1980s when catch rates
fell worldwide. Ransom and Boris end on a laconic note: "Industrial-
ized fisheries typically reduced community biomass by 80% within 15
years of exploitation." "We estimate," they add, "that large predatory
fish biomass today is only about 10% of pre-industrial levels."

Their article[7] made the cover of *Nature* and, according to Boris,
"Ransom spent the next few weeks answering so many calls from journalists that he lost his voice." In France, *Le Monde* ran the headline
"Large fish stocks in danger of disappearing," calling it a "shocking
realization,"[8] and the Japanese administration kicked themselves for
letting the Canadian team peek at their reports. "Of course, everyone came after us, especially the fisheries management organizations,"
Boris sighs. "Daniel was the only one who was a good sport—he even
supported us in the press." The younger scientist steadied his nerves by
telling himself, "They're criticizing my work, not me personally." Ransom, on the other hand, was more belligerent: "They're wrong, and I
am right!" he declared.[9] But he was distressed when he read an article
written by his colleague (and supposed friend) Carl Walters titled "Folly
and Fantasy in the Analysis of Spatial Catch Rate Data."[10] "Ransom
was so torn up about it that he didn't even tell me," Boris recalls. "Carl
went a little overboard when it came to the tone."†

The experience was painful, but Myers and Worm had finally
torn a hole in the wall of indifference that couldn't be ignored: "If

---

* The quote was initially reported by Nancy Baron.
† According to the German expression Boris Worm used in our interview (June 16,
2017): "Carl hat sich im Ton ein bißchen vergriffen."

fisheries conservation biology and its guiding philosophy thrive, it will be because of the energies of the likes of Ransom," Daniel wrote several years later.[11]

THE RESULTS OF the Sea Around Us project, Myers and Worm's analyses, and the work of several other teams from around the world proved most instructive.

Joshua Reichert, bringing the project full circle, asked Daniel to go beyond his usual reports and scientific publications—if the news was going to reach decision makers, they had to publish a book for the general public, and fast. Nothing excites Daniel more than the idea of diving headfirst into another wordsmithing spree, but he simply didn't have the time, so he called for backup in the form of Jay Maclean. "Daniel came up to me and said, 'I want you to write this book, here is the money, I know you'll be wanting it for your son's education,'" Jay recalls.

A worthy successor of the illustrious Rachel Carson, Jay got to work gathering his sources, including the essential *Sea of Slaughter* by Farley Mowat. Since library catalogs in the Philippines were somewhat limited, Jay spent several weeks in Rome, locked up in the FAO archives. Back at his home in Anilao with its view of the coral reef, he began crafting the simple, precise, and elegantly turned phrases he had become known for. Jay fulfilled his contract, delivering about a hundred pages accompanied by a generous helping of notes. Reichert and the publisher came back with two requests. First, they asked that the book be signed Pauly and Maclean, arguing that it would be easier to sell with the better-known author listed first. Mightily embarrassed, Daniel passed the publisher's demands on to his old friend. But Jay only shrugged—he was used to being a ghostwriter. The publisher's second request was that Daniel write the introduction himself, a project he completed during a stay in the French countryside.

In fact, when Daniel's adoptive father, Louis Pauly (or Loulou, as he was often called), retired from his job in heavy industry in the late 1970s, he couldn't wait to escape the gray skies and government housing in the Parisian suburb of Maisons-Alfort,* and his wife, Renée, couldn't agree more. But their children were spread out all over the country, except for Daniel, who had just left for the Philippines. Ever the rationalist, Renée spread a map of France on the kitchen table, and a few barycentric calculations later, she had the answer: La Creuse, a rural region smack in the middle of the hexagon. They found a house within their budget—a two-room cottage with a large garden near the hamlet of La Vallade, about twelve miles northwest of Guéret. Loulou moved in right away, only too happy to start his vegetable garden. Renée joined him a few years later, leaving her department store job far behind her. The home was rustic—at first, there wasn't even running water or central heating. But the Paulys fixed it up and expanded it over the years with help from their children. "They were exhausted from all those years of work," one of their neighbors told me. "Loulou didn't live to be very old. You could tell they hadn't been very happy in Paris, but here, they were." Renée and Loulou had never gotten their driver's licenses, so they crisscrossed the countryside on their mopeds (they each had their own—Renée had never needed a man to tell her where to go), doing their shopping in the nearest village, a few miles away. Monsieur went to buy his cigarettes, Madame to load up on books, which she devoured greedily before narrating their content, in detail, to anyone who was willing to listen.

For the next thirty years, La Vallade became the rallying point for the whole Pauly clan. In order to house everyone in the summertime, they put up tents in the garden and converted the attic into a sleeping loft. The river was right around the corner, a good place to swim, surrounded by greenery and with a sandy beach, just like at the seaside. In the summer of 1981, Daniel, Sandra, Ilya, and baby Angela visited from Manila. A Super 8 home video shows them sitting peacefully among the younger Pauly brothers, who horse around for the camera. "La Creuse helped me make my peace with France," reminisces Daniel,

---

* The Parisian suburbs ("la banlieue parisienne") carry a negative stigma akin to that of the "inner city" in the United States. (TN)

who enjoyed the beautiful countryside and talked with his mother from morning till night. He returned to La Creuse again and again, not as often as he would have liked, but anytime his mad dashes around the world led him close enough to Paris.

And that is precisely what happened in 2002, shortly after his publisher asked him to write an introduction for Jay's book. "I was attending a UNESCO workshop in Paris and it was so idiotic that I just left," Daniel recalls. "I rented a car, and I went to see my mother. I was a little depressed by everything I'd seen in Paris and some other things as well. In La Vallade, I wrote the introduction in a sort of trance, describing an ocean where we'd wiped everything out, where there was nothing left but old beer cans rolling back and forth on the sea floor."

DANIEL'S TEXT IS poetic, as authoritative and enlightening as one of Rachel Carson's essays. He begins with the wreck of the *Andrea Gail*, a 72-foot fishing vessel that left Gloucester, near Boston, in September of 1991 to chase swordfish, only to disappear thirty-eight days later off the coast of Newfoundland beneath 60-foot-high waves and winds over 100 miles per hour. It was the kind of story that was made for the movies, and Hollywood, true to form, glorified the epic battle between man and the "merciless" ocean.* For Daniel, though, the story lay elsewhere: What was a relatively small fishing boat doing at the end of October more than 600 miles off the Canadian coast? Running dangerously low on fuel, the *Andrea Gail* was in fact hunting down the few remaining swordfish in the North Atlantic—it was overfishing that killed those six sailors. After that dramatic introduction, Jay and Daniel pick up the sad narrative of the pillaging of the North Atlantic, supported by the Sea Around Us's data on the twentieth century. Having successfully painted a portrait of the current state of affairs, the duo asks the question: "How did we get here?"

---

* Wolfgang Petersen, *The Perfect Storm*, 2000, 130'. The film is based on the book of the same title by Sebastian Junger, published in 1997.

"There are no villains, least of all the fishers," they stress from the outset.[12] For them, resource management in the region simply didn't correspond with ecological reality. The root of the problem is likely conceptual: fisheries biology operates on the myth of infinite ocean resources and manages fish stocks species by species without taking fishing's ecosystem-wide effects into account. Additionally, even as the profitability of ocean fisheries is diminishing, fishing efforts are being increased thanks to subsidies from national governments, around thirty billion dollars per year worldwide—a mind-boggling sum. In practice, this means that 16 percent of the price of any given seafood item in Europe is covered by subsidies, a figure that climbs as high as 25 percent in Canada. These figures are available thanks to the work of Rashid Sumaila and his very determined team, who wrested them from the official and semi-official documents of different North Atlantic countries.

Indeed, the synthesis published by the Sea Around Us launched Rashid Sumaila's career. A young Nigerian economist trained in Norway, he moved to Vancouver in the mid-nineties and quickly became a world-class specialist in the socioeconomics of fisheries, eventually landing a professorship at UBC. "I like Rashid a lot," says Daniel (who is rarely so effusive) when I ask about his young colleague. "And I respect him. I think the feeling is mutual." Certainly, when I finally meet Rashid in Vancouver, he strikes me as the friendliest and most good-humored member of his department. It is Daniel who tells me about Rashid's early career: "In Norway, he worked so hard and he was so productive that he had to leave—it's what's called 'tall poppy syndrome.'* It was Tony Pitcher who invited him to Vancouver, and I quickly realized that he had a great work ethic. We had contests to see who could work the longest each day—I lost."

The master advised his student to look at the big picture, and Rashid did just that, studying the financing and socioeconomic impacts of fishing, not for a specific part of the Norwegian or West African

---

* Tall poppy syndrome describes a situation in which individuals perceived as more successful than their peers are resented or attacked.

coast, but on a global scale. Perhaps most remarkably, he created the first international databases of ex-vessel seafood prices* and the total cost of fishing activities and their socioeconomic benefits, as well as of fishing subsidies allotted by national governments. "Those projects got him all the way to the World Bank and the World Trade Organization," Daniel remarks proudly. When Sumaila presented his findings in Brussels, Daniel couldn't help but smile at the sight of an African explaining to Europeans how to run their fisheries. In particular, Sumaila and his colleagues criticized the government subsidies allotted to fisheries, pointing out that they promote industrial fishing at the expense of more local, environmentally friendly, and socioeconomically sustainable fisheries. Essentially, all of the fisheries management strategies already in place were based on short-term goals without taking the needs of future generations into account.

The economic model being applied to North Atlantic fisheries, among others, was therefore not viable in the long term, and worse, the existing systems of governance were ineffective. "We have not used the regulatory tools and developed the right institutions for managing fisheries effectively for their long-term sustenance," Jay and Daniel contend in their book. They go on to provide a list of sixteen commissions and forty-three international mechanisms that regulate fishing in the North Atlantic—though with mixed results. The Northern European countries make an honest effort, but Southern Europe is a different matter. The European Commission sets fishing quotas but leaves it up to its member states to make sure that those quotas are respected within their borders. Some countries—such as France, Greece, and Spain—tend to dig in their heels: they would rather pay tens of millions of euros in fines to the EC than make their fishers respect the quotas. Overall, the means allotted to keeping tabs on fishing activities are also largely insufficient, even in countries like the United States and Canada, where the size of areas being exploited is quickly increasing as a result of overfishing. Jay and Daniel conclude that "just as in

---

* Ex-vessel prices: prices at the landing site, i.e., at first sale. (TN)

the case of control of vehicular traffic by police, control by fisheries officers cannot fully replace voluntary compliance by the majority,"[13] adding: "All users of marine ecosystems... take part in the ocean's stewardship... Status quo is not enough. Restoration of past levels of abundance is required."[14]

For that to happen, Jay and Daniel conclude, we must set aside the current approach, which focuses exclusively on species exploited by fisheries, and work to restore the ecosystems on which those species depend. Thus the idea of ecosystem-based fisheries management was born. During the first few years of the twenty-first century, Daniel's publications and those of numerous colleagues helped develop this notion, which they define as follows: "The intent of the concept of marine ecosystem management is to place ecosystem integrity or health as the primary consideration in all management decisions that affect the ecosystem. From an economic viewpoint, the ecosystem restoration and management can be seen as an investment in a natural resource."[15] They also suggested a list of five urgent measures designed to mend ecosystems in a North Atlantic damaged by fisheries. They understood that in doing so, they were going beyond their role as simple ecologists and leaning dangerously on "the firewall between science and advocacy." Because their proposals perfectly summarize the objectives that need to be attained if we are to restore the oceans, I have chosen to quote them at length:

> Fishing pressure on North Atlantic fish populations and ecosystems must be drastically reduced, by a factor of three to four in most areas... One very effective tool, in this context, is the abolition of subsidies... Large marine reserves, amounting to at least 20% of the ocean by the year 2020, must be established, pending the long-term transition toward a regime where specific areas are explicitly open for fishing, while the rest is closed by default.
>
> Eco-labeling and other market-based efforts to move the fishing industry toward sustainable practices must be intensified. An effective regime must be designed and implemented to publicly

expose deliberately unsustainable and illegal practices, and their perpetrators.

Access and property rights in fisheries should favor smaller scale, place-based operations, operating passive gear to the extent possible, and the fisheries should be run through co-management arrangements.

When they say "passive gear," Daniel and Jay are referring to traps and other fishing techniques that are far more selective and less destructive than trawling, a method that can transform rich ocean bottoms into lifeless deserts. They add that they are not defending artisanal fisheries for "romantic" reasons, but rather because "local fishers, if given privileged access, will tend to avoid trashing their local stocks, while foreign fishers do not have such motivation."

The book, signed Daniel Pauly and Jay Maclean, *In a Perfect Ocean: The State of Fisheries and Ecosystems in the North Atlantic Ocean*, was released in 2003 by Island Press, an American publisher specializing in environmental literature. At 175 pages, it reads like a detective novel and resonates like an alarm, though the text is also infused with Jay and Daniel's usual dry humor. "A few more decades of long-lining will resolve the problem, as extinct sharks cannot be finned and discarded," they quip at one point. Or later: "You cannot go lower than that in terms of fishing down marine food webs: sea cucumbers eat dirt." I read the book during an expedition to Greenland in the summer of 2004, and I could barely put it down. Back in France, I contacted Daniel for the first time, sending him an email that proved to be the first of many.

CHINESE FISHERIES
AND CHARLES DARWIN

"I SMELLED A RAT," Daniel declared.[1] He had studied the North Atlantic at the request of the Pew Charitable Trusts, but his end goal with the Sea Around Us was to study fisheries on a global scale. Starting in 2000, he began examining FAO catch data on global fisheries more closely. Notwithstanding all his recent findings on the collapse of predatory fish stocks and fishing down marine food webs, the annual volume caught seemed to keep growing throughout the 1990s. In 1993, it even exceeded the symbolic limit of 100 million metric tons annually, according to the FAO. Still, pro-fishing lobbies argued repeatedly that the ocean's resources were infinite, claiming that things were going swimmingly beneath the waves.

The Sea Around Us team was skeptical, however, and in April of 2001, Daniel mentioned his concerns to the FAO executives who had come to audit the Vancouver group's work for the Pew Charitable Trusts. In particular, he spoke with Richard Grainger, then head of the FAO statistics department, about how the earliest world catch maps generated by Reg Watson indicated insanely high catches for Chinese waters, despite the fact that they had long been classified as overfished. "Grainger smiled knowingly: everybody in this group knew that China had been overreporting its marine catch for years," Daniel recalls.[2] He asked Grainger why the FAO didn't simply right the ship: Grainger replied that modifying statistics supplied by member states was not part of their mission. After some discussion, Grainger did, however, "order" a study from Reg and Daniel that would estimate how much China overdeclared its catch, with the end goal of taking those figures into account. Reg Watson got to work: thanks to the information in FishBase on the species present in Chinese waters and some statistical models that allowed him to estimate their productivity, he managed to calculate how much the Chinese *should* have been able to catch. For 1999, it turned out to be half of what China had reported to the FAO. Ultimately, after Reg corrected the fisheries statistics from 1970 to 2000, it became obvious that global fisheries were not expanding—in fact, they had been in constant decline since 1988.

But why would there be so much statistical inflation coming out of China, when in the rest of the world, countries tend to underestimate their catch in order to cover up latent overfishing? To answer that question, Daniel recruited Lei Pang, a Chinese scientist well versed in geopolitics who had recently immigrated to Canada. Pang identified and translated numerous texts from Chinese websites. Her detective work showed that the central government was well aware of the problem of "watery statistics" (as they're called in Chinese), an issue rooted in communist political culture. In China, promotions for administrative heads are largely based on growth in the sector they supervise, just like in Stakhanov's Russia.* This distortion affects every sector of the Chinese economy. The theoretical energy needs of the country according to the "watery statistics," for example, is twice the total output of all its power plants.

Reg Watson, Daniel Pauly, and Lei Pang wrote two reports based on their findings and sent them off to the FAO in the summer of 2001. "Where it would have remained accessible only to experts," Daniel points out. "Thinking this would be interesting to a wider audience," he continues, "Reg Watson and I also summarized the report into a shorter contribution, which we submitted to *Nature*."[3] It should be

---

* Russian miner Aleksey Stakhanov was the founder of the Stakhanovite movement, a Soviet workers' movement that began in the 1930s and that aimed to increase worker productivity in order to demonstrate the superiority of the socialist economic system.

noted that the Sea Around Us team did not tell the FAO what they were doing. "It might have crossed the line," Daniel admits, "but since the data was accessible to everyone already, and we had already sent our results to the FAO, I don't think we were at fault ethically." Yet again, Daniel's article made it past the firewall of editors and reviewers and was published by *Nature* in November of 2001.[4]

"And that's when the shit hit the fan," Daniel recalls. *Nature*, sensing a potentially newsworthy controversy, had published the article along-side an editorial that blasted the FAO and a political cartoon that poked fun at both that venerable institution and the Chinese. After that, the FAO was ridiculed in the press for its inflexible policies, while China took flak for its outmoded administration. The media storm was all the more intense because, for the first time, Daniel called in a professional to write a press release and send it out shortly before the article appeared in *Nature*. He chose Nancy Baron, whom he'd met after his first publication in *Science* in 1998. She scrubbed the scientific jargon from their text and personally contacted dozens of science journalists to explain the study's goals and geopolitical context. Three months after the *Nature* article was published, Nancy proudly presented Daniel with nearly one hundred newspaper articles (including two in the *New York Times*) and radio shows, mainly in North America, Europe, and... China. "This was especially remarkable," Daniel points out, "given the shrinking space for science journalism following the September 11 tragedy."[5]

Meanwhile at the FAO, Richard Grainger and his team seethed. Oddly enough, though, they did not reveal to the press that they were the ones who had ordered Watson and Pauly's study—but they did throw the Sea Around Us team's reports out the window and instead published a long-winded response on the FAO website. The document did not call into question the essential fact that certain fishing-related statistics supplied by national governments are fatally inaccurate, how-ever. The Chinese response, on the other hand, was more ambivalent. The Beijing bureaucrats initially admitted the problem but, likely embarrassed by all the media coverage, soon retreated into denial, a position relayed by the state-run press agency. "We received some nasty emails, which tended to elaborate on the official position, as reported by the Xinhua News Agency," Daniel recounts.[6] "The saddest of these private communications was from the director of a biochemistry laboratory, who wrote that the reason Chinese fishermen have 'big catches' is because they work hard, as opposed to Africans who don't, etc. Clearly, we still have a long way to go."[7] Chinese reactions from outside the government were more nuanced, but the Taiwanese media jumped at the chance to torment their communist brothers—a particularly provocative article in the *Liberty Times* bore the title "China, a Nation Good at Statistical Falsification."

"As the preceding exchanges illustrate," Daniel concludes, "the relationship between science and policy (and politics!) can be very tense. In the case of the FAO, politics clearly trumped science, not out of malice on the part of the scientists, of course. Most fisheries scientists—and probably most other scientists working on natural resource exploitation—work in government agencies and are expected to remain silent, even in the face of obvious conflicts of interest." Daniel points out that it is therefore university-based scientists who must take up the important responsibility of keeping the public informed, because they are often able to speak more freely. "The public's need to know," he insists, "also extends to 'scientific' matters."[8]

The Chinese fisheries incident proved disastrous for the already strained relations between Daniel and the FAO, but it also cemented his international reputation. He had become indispensable: soon, *Nature*[9] and then *Science*[10] both commissioned synthesis reports on the sustainable development of fisheries and their future, which gave Daniel a space to develop his own ideas about the current state of marine resources and how they could be used to build a better world.

DANIEL'S FIRST TEN years in Vancouver forged his international career and his reputation as a whistleblower. Away from the spotlight, however, he was leading a double life, pursuing a captivating, secret, and all-consuming affair. His enchanting and domineering muse? None other than Charles Darwin, for whom Daniel nursed an intense passion.

It all started in Peru in the late eighties, while Daniel was working on the sequel to his book on the Humboldt Current. He remembered that Charles Darwin had passed through the region during his famous trip around the world aboard the *Beagle* between December 1831 and October 1836. Darwin had collected a few anchoveta specimens that would later be described under the scientific name *Engraulis ringens* and identified as the keystone of the coastal marine ecosystem of the South American coast, from Ecuador to Chile. Daniel decided to spruce up his volume with an epigraph from Darwin and began poring over everything the man had written. He never found the right quote— Darwin actually wasn't that interested in the anchoveta—but his foray into this long-dead scientist's world proved to be a voyage of no return.

Plunging into the account of the *Beagle*'s voyage, he discovered a sympathetic and perceptive Darwin who, though he hadn't yet formulated his theory of evolution, spoke intelligently about current events and the major issues of his time. "I realized that Darwin was an abolitionist and that many of his biographers had ignored that side of him despite the fact that he developed his ideas on the subject at length in *The Voyage of the Beagle*. Of course, my reading of Darwin reflects my own obsessions, but it's unbelievable that the fight against slavery is never mentioned in conjunction with the name Charles Darwin, even though his whole family was abolitionist." During his studies in Edinburgh, the young Darwin learned the art of taxidermy from John Edmonstone, an African born into slavery in British Guiana. "I used often to sit with him, for he was a very pleasant and intelligent man," wrote Darwin, who was unsurprisingly sickened by the colonialist, slaveholding societies he encountered in South America. Much later, in 1871, Darwin published *The Descent of Man, and Selection in Relation to Sex*, a treatise that traced the common evolutionary history of every human being on the planet at a time when his European contemporaries were using racial supremacy to justify their colonial policies. "Of course, Darwin's ideas fascinated me," Daniel tells me, "but as I read on, I got to know him as an extraordinarily sensitive and likable human being. That's not often the way with famous men—usually, when you get beneath the surface, you discover a monster."

Daniel began writing, in his way, a declaration of love for Charles Darwin: he went looking for every sentence Darwin had ever written about fish and published them all in a volume called *Darwin's Fishes: An Encyclopedia of Ichthyology, Ecology, and Evolution*. His starting point was, of course, the resemblance between "Darwin's finches" and "Darwin's fishes." Indeed, it is often said that the different species of finch he saw in the Galápagos sparked the idea of the evolution of species. Though this isn't strictly true, "Darwin's finches" has become a popular saying, and the evolution of the little feathered creatures is still the subject of extensive research today.[11] Daniel—initially as a joke—decided that Darwin's fishes were just as important as his finches. Over several years, he read everything Darwin had ever written—over six million words in twenty-nine volumes, plus copious correspondence.[12] Darwin became his refuge, the personal project that he could develop at his own pace without having to worry about an army of partners whose shortcomings and delays often drove him to distraction. He worked in his secret garden early in the morning and on weekends, especially during his first few solitary years in Vancouver. The dock-front cafés near downtown became his favorite haunt. "I would eat brunch while reading the *New York Times* Sunday edition. Once I was well-fed and well-informed, I would tackle Darwin."

Ultimately, Daniel collected 45,000 of Darwin's words on fish and put them together in the form of a *"dictionnaire amoureux,"** accompanying each entry with a commentary of his own devising. It took him seven years to finish the first, nearly complete, version. In 2000, he presented the results of his labor to a few colleagues during a conference

---

* *Dictionnaire amoureux*: A literary genre invented in France in the early twenty-first century, a *dictionnaire amoureux* is made up of a series of short articles, usually arranged in alphabetical order (hence the name *dictionnaire*, or "dictionary"). They are not designed to be reference volumes, but rather collections of subjective essays on a given subject (hence the adjective *amoureux*, "in love"). (TN)

246 | THE OCEAN'S WHISTLEBLOWER

held, quite appropriately, at the Charles Darwin Research Station in the Galápagos. "We were blown away, especially because Daniel had accomplished such a huge task alongside his normal research," recalls Coleen Moloney, who was present for the conference.

Next, Daniel used his growing notoriety to convince an editor to publish his somewhat eccentric project. The monument Daniel had built to Darwin's glory finally saw the light of day in 2004. Its preface, written in typical Paulyesque style, deserves to be cited at length: "Many sections of this book read like laundry lists," he writes. "I have attempted to cover this up, mainly through levity, the result being that this book will probably irritate serious scholars, but still bore students to tears." And further on: "I have emphasized the few errors I found, both because Charles Darwin is such a worthy target, and because without such emphasis, this book may be perceived as a hagiography."[13] Joseph Nelson of the University of Alberta described it as "an adventure in learning." I ask Daniel if it sold well: "I prefer not to know," he sighs, "it's too depressing."

Daniel's book about Darwin was not destined to become a bestseller. He also continued to be frustrated by the indifference of his colleagues toward his favorite hypothesis about the surface area of fish gills limiting the fishes' size—a subject on which he still published regularly. Unlike everyone else, he considered his work on the environmental impact of fisheries to be nothing more than an exercise in methodology. These were, however, the advances that would see him covered in glory—even if they were not sufficiently conceptual for his own taste.

UNCERTAIN GLORY

"DR. PAULY HAS been alternately described by colleagues as inspiring, arrogant, brilliant and aggravating. Some call him a heretic, 'the Prophet Daniel.'"[1]

"For better or worse, he's probably had a greater impact on the field than any member of his generation."[2]

Though there was some gnashing of teeth, Daniel's career had reached its apogee. His articles and those of the Sea Around Us were being published in the best scientific journals on the planet and cited thousands of times, though this certainly didn't stop him from authoring and coauthoring a dozen more papers, book chapters, and reports for less prestigious outlets each year. Always a generous writer, Daniel would produce a text for anyone who asked politely for a few pages on fish and the state of the world's fisheries. As his beloved Darwin put it, his mind "seemed to have become a kind of machine for grinding general laws out of large collections of facts."[3] But unlike Darwin, who was literally nauseated by public speaking, Daniel excelled at treading the boards. "You have to get the audience's attention as quickly as possible, just like an actor," he explained to me. "Of course, I get stage fright, but after the first couple of minutes, if the public is with you, it's an immense pleasure." Invited to speak all over the world, he began appearing at about fifteen conferences a year, rarely refusing an opportunity to travel. "When I get on a plane," he jokes, "I turn into a cold-blooded animal. My metabolism slows down, and I just wait till it's over. That's how I'm able to stand the inevitable delays, detours, and other unpleasantness." At international conferences, no one wanted to go on after him, other speakers dreaded his questions, and they walked on eggshells around him during meetings.

The media gradually discovered some of the details of his pica-resque early life alongside his scientific exploits. Several long profiles followed, including one in his beloved *New York Times* in 2003. That same year, *Nature* published his responses to a science-themed Proust Questionnaire.[4] Some of his answers proved revealing:

Summarize yourself in the form of a title of a paper in Nature.
Longitudinal single case study documents instance of a scientist "making it" notwithstanding apparent lack of key skills.

What book has been most influential in your scientific career?
Leftist political tracts swirling about in the 1960s, which convinced me that I should study something that would "serve the people," rather than indulge in my literary, historical and philosophical interests. As it turned out, I was able to do it all.

Where and when would you most like to have lived or worked?
Subtract a few decades and I, as a black person, could not have been an academic anywhere. So I much prefer the present.

What music would you have played at your funeral?
Aretha Franklin's "R.E.S.P.E.C.T."

How would you like to be remembered?
As the one who showed that the effect of fisheries on marine life is equivalent to that of a large meteor strike on terrestrial life.

What's just around the corner?
A representation of global ocean ecosystems that we immodestly call "World Model," contrasting the structure of present marine ecosystems, and their fish biomass, with those of the 1950s. This should help to settle the issue of fisheries impacts, as well as

providing a solid framework for discussions about future global fish supply as required, for example, by the United Nations-sponsored Millennium Ecosystem Assessment.

In 2003, Daniel also became director of the UBC Fisheries Centre in Vancouver, a responsibility that he accepted "as a final resort, just to make sure someone horrible didn't end up with the job." With a little over a million dollars a year, the Pew grant allowed him to enlarge the staff of the Sea Around Us. In particular, he prided himself on being able to direct ten doctoral students at once. With his research assistants, Daniel remained as charming as ever, dogmatic but with a sense of humor, just like in the old days in Manila. With his PhD students, though, it was another story. Daniel set the bar astronomically high, and, considering his own capacity for work and ease when it comes to writing, he has never understood why his students don't advance at lighting speed. According to most of his colleagues, it is difficult, perhaps impossible, for those young researchers to exist intellectually next to Daniel Pauly. As a result, his students after 2000 don't always have a good relationship with their maestro, but it is telling that they all acknowledge his influence on their work. His team is, as usual, both international and mostly female. "I usually get along better with women. Young men, sooner or later, always feel the need to test me," he tells me with a glint of humor in his eyes. Daniel almost never makes hires based on a resumé, relying mostly on chance meetings and recommendations from his considerable network of associates.

This network and the contacts he maintains in different places around the globe have earned him eight honorary university degrees. The first was, logically, an honorary professorship from IfM in Kiel. The second, somewhat surprisingly, came from the Aristotle University of Thessaloniki. The man behind this distinction, Kostas Stergiou, is director of the Hellenic Centre for Marine Research. According to Kostas's younger colleagues, "When Daniel comes to Greece, he's greeted like a rock star." Daniel's Mediterranean alter ego is an elegant, white-haired man with sea-blue eyes. In his rough smoker's voice, Kostas tells me about their first meeting:

> I met Daniel Pauly for the first time in Athens in 1992, during the first World Fisheries Congress. We had a long discussion in the shadow of the Acropolis—over juice (him, he is a teetotaler) and ouzo (me)—on various scientific issues. The next year, when we spent a week together in Monterey, California... we had the opportunity to extensively discuss various issues; one thing I remember from our discussions is what he told me at one point, "We will grow old together..." Back then, Daniel was a very charismatic speaker, enthusiastic, motivated, revolutionary, persistent (he never lets his ideas go unnoticed)... He is also an unconditional friend. He likes movies and TV series a lot, and during one of my visits to Vancouver I set up a home cinema for him.

DANIEL SURFED ON a wave of success, but an important ordeal still awaited him. Usually cheerful, his voice grows serious as he tells me this story: "It was in late January 2005; I'd been working like crazy all week, sending in a draft of one article and editing another." Sandra, now supervising an international baccalaureate program, spent the day at a conference, then went to dinner in downtown Vancouver with some coworkers. She invited her husband and Angela, who was living with her parents while she completed her master's degree at UBC, to come with them. "I was so overworked and on edge that I felt like a runaway train," Daniel remembers. He parked downtown and went for a walk alone to calm his nerves. "I was in sorry shape," he admits.

He made it to the restaurant late, but still in time to see Sandra's colleagues. "Daniel was quite popular at those events," she tells me. "That night, he quickly got into a very animated discussion with one of my French colleagues—but suddenly, he told us it was time to go. Angela and I both thought that was strange." Sandra offered to drive, but Daniel refused. Their home on the UBC campus was only fifteen minutes away. "While I was driving, I started to feel strange," Daniel recalls. "When we got back, I had all the symptoms of a stroke, which was affecting my right side, but I could still stand and I decided to just go to bed. If I'd been in my right mind, if I'd been able to better evaluate the situation, I would have gone straight to the university hospital, just around the corner, where they would have given me anticoagulants. But I went to bed and the next day, it was too late." Daniel woke the next morning and pretended everything was fine. Despite falling a couple of times in the bathroom, he managed to get dressed. At breakfast, though, he was finally exposed—he couldn't hold his knife. "We were terrified," Angela tells me. "He didn't want to hear anything about it," Sandra adds, "but we called my sister Cheryl in California." Cheryl is a psychologist with a good knowledge of medicine. She is also, as Angela notes, "the only person who can get Daniel to budge at times like that."

"When we arrived at the hospital, there was nothing to be done, really," Daniel tells me with a hitch in his voice. "The brain tissue around the clot was already dead." Daniel suffered some facial paralysis, which disappeared after the first week. Still, he had lost the use of his right arm and had a limp on the same side. "I was supposed to leave for England that day to meet Prince Charles. It's ridiculous, but I felt so guilty about canceling." Daniel was also terrified at the idea that his mind may have been affected: "When they were doing the first tests, they asked me to name twenty animals and I forced myself to remember the names of twenty dinosaurs."

He put everything on hold for a whole month, except for a book review, which he wrote the day after his stroke just to assure himself that he wasn't "a vegetable." "It was one of his best reviews, actually," insists Sandra. He also spoke regularly to the journalist Bruce Knecht, who called him seeking advice for his hard-hitting book on the illegal exploitation of Patagonian toothfish (often marketed as Chilean sea bass), a species recently discovered by deep-ocean trawlers.[5] But most of the time, Daniel was completely exhausted. For the first time since he lived in Kiel with Sandra, he watched TV, finally discovering the strange and wonderful series that so fascinate the rest of humanity. Meanwhile, Daniel and Sandra received a flood of supportive messages.

"I especially liked the one from Rainer Froese," Sandra relates. "He titled it *Austeigen verboten*."* In the scientific community, news of Daniel's stroke spread like wildfire, inspiring a lot of sympathy, but also a few knowing smiles: they might finally be rid of that troublemaker Daniel Pauly.

He recovered quickly, though, thanks to a series of in-home treatments during which needles were plunged deep into the muscles of his shoulders and back. "It took about a dozen sessions, four or five per month, each with fifty needles or so," explains Daniel, without mentioning the pain caused by these porcupine attacks. He gradually regained the use of his right hand, then his elbow, then his whole arm, though not his fine motor skills. "It does make me angry that Daniel didn't take the time to finish his rehab," Sandra admits. "It would have improved his physical condition significantly. But Daniel wasn't interested—he just kept working, as usual."

"I still feel the effects some," Daniel admits. "I can't whistle. I can't type with both hands like I learned to do at school in La Chaux-de-Fonds. I can't jump. When I am tired, I have a slight limp." Daniel had never been so happy to be left-handed and mentally thanked his Swiss schoolteachers for not forcing him to switch. He returned to the office only three months after the incident and began traveling again by June of 2005. His family and friends worried about him at first. "It drove me crazy," says the usually imperturbable Sandra. Roger Pullin recalls, "That summer, I saw Daniel at a conference in the Netherlands. During his presentation, he fell off the stage, but I just thought it was because of his natural enthusiasm." Jay Maclean received his convalescent friend at his home on the Philippine coast: "He insisted on going snorkeling. It started alright, but he just could *not* get back. He nearly drowned me while I was helping him back to shore."

In the end, though, everyone I spoke to agreed on one point: a new Daniel was born in 2005, a more sensitive one, a better listener—even

---

* Literally, "Do not exit while train is in motion." A figurative expression in German that means "Don't leave before the party's over" or "Don't leave without saying goodbye."

if his relentless dedication to his work remained unaffected. For his part, Daniel takes a rational approach: "When you have a stroke, it changes the way your brain works and affects your hormones, lowering the threshold for emotiveness. You laugh and cry more easily. What really gets to me now are stories about altruism—I even tear up reading the newspaper now because I'm so overwhelmed by emotion." Daniel was condemned to a lifetime aspirin regimen, but it's a small price to pay—he hasn't had another stroke since. And most importantly, Daniel's stroke allowed Sandra to get her husband back. The spouses had been living separate lives for too long. Daniel finally let his guard down and admitted how much he owes to his wife of thirty years—he is nothing without her. And Sandra, who retired soon after, has been more involved in her husband's daily life, often going with him on trips and keeping an eye on his workload: "Daniel doesn't know how to say 'no' and a lot of people take advantage of him," she says.

Daniel had (almost) worked himself to death: running the media gauntlet, exposing himself to criticism from the pro-fishing lobby, and receiving threatening and insulting letters from fishers on the East Coast of the United States, among other ordeals. I wonder to what extent these pressures affected his health. When talking to journalists, Daniel has always claimed that criticism doesn't affect him, but in private, he admits that all the attacks and small betrayals often got under his skin.

His "friendly competition," Boris Worm, who courageously withstood the media attention of the early 2000s, also spoke to me about that time. Boris, at the head of a team of thirteen other scientists, published a study in *Science* on November 3, 2006, that made a great deal of noise. The paper analyzed the impact of the loss of marine biodiversity on ecosystem services. They found, unsurprisingly, that the state of marine biodiversity was not a happy one, but that at this stage, things could still be turned around: with a little effort and a strong political will, marine ecosystems could be restored and food security ensured for all. Boris wanted to adopt a more conciliatory and less apocalyptic stance than the older generation of scientists like Daniel or his friend

Jeremy Jackson, who studies historical overfishing. Buried among the conclusions of Boris and his team, though, was a serious warning: if nothing is done, the world's exploited fish stocks will all disappear by 2048. The article was a hit with the media thanks to a press campaign organized by the formidable Nancy Baron, who worked simultaneously for Daniel and the Halifax team, but also because of how *Science* pitched the article. "They put the disappearance of exploited fish populations by 2048 in the forefront," sighs Boris, "and the media focused on that a lot more than we would have liked."

His colleagues and the fishing sector called Boris all sorts of names, accusing him of "talking nonsense" and undermining the hard work of fisheries specialists who are managing stocks as well as they possibly can. Soon, though, Boris had bigger things to worry about. A few weeks after the article appeared in *Science*, Ransom Myers was hospitalized with an inoperable brain tumor. He died five weeks later, leaving behind his wife and five children. "Heike [Lotze, Boris's partner at the time] and I found ourselves alone, trying to run the research empire that Ransom had built in Halifax—I had only been on the job for eighteen months and it was just too much." Boris fell ill with cancer the following year, but his sunny personality and an abundant pharmacopeia saved his life: "I happened to have a great medical team that I trusted completely." He survived and went on publishing high-level scientific articles with insulting frequency. He remained a media darling, and he and Heike Lotze were crowned the most glamorous couple in marine ecology.[6]

When I ask about the dark years following Ransom Myers's death and the pressure that Daniel and other whistleblowers were under, Boris answers simply, "Stress is obviously an aggravating factor for all kinds of pathologies, including cancer—it's been proven scientifically, right?"

Daniel had a close call, but his stroke barely impacted his career. Indeed, starting in the early 2000s, the honors just kept flooding in, making him the most decorated marine biologist in the history of the discipline. The shower of awards proved particularly heavy in the United States, where *Scientific American* named the abandoned son of

Winston McLemore as one of the most influential scientists on the planet. In 2003, he joined the Canadian Academy of Sciences. His native France was a little behind the times, but Daniel, the former draft dodger, was awarded the Grand Prix of the Société française d'écologie in 2011 and the Légion d'honneur in 2017. The most unexpected accolade came from Japan. "In June of 2005, I was in France to evaluate the fisheries research center in Sète," Daniel recalls, "and I received an email telling me that I'd won an insane amount of yen." Daniel transferred the suspicious message to his spam folder, but then, thinking the better of it, did some research: he had in fact just won the International Cosmos Prize, awarded each year by a Japanese association for outstanding work promoting a harmonious existence between nature and mankind. Past laureates had included the iconoclastic atheist biologist Richard Dawkins and the writer and scientist Jared Diamond. A few years later, Japan honored Edward O. Wilson, the champion of sociobiology and biodiversity, then French anthropologist Philippe Descola, and finally, in 2017, the primatologist Jane Goodall. In light of Daniel's unflattering comments about Japanese whalers and long-liners, such a distinction from one of the world's fishing capitals came as quite a surprise. "I learned that Nancy Knowlton, who is a coral reef specialist, was on the selection committee, and I think she was probably the one who suggested me," Daniel explains. "It was a courageous decision on the part of the jury, which was mostly Japanese."

After the initial decision, though, a more "political" committee took charge: "They had sent letters to quite a few of my colleagues asking if I was the kind of person who would cause a problem, be too provocative, et cetera." His nomination made it through the red tape— no one knows how—and they sent him the good news. "That day in Sète, Daniel took us all out to a restaurant to celebrate," remembers Philippe Cury, director of the fisheries research center at the time. "An FAO representative was also there, and he was just green with jealousy!"

"But that's not all," Daniel adds with a laugh. "The Cosmos Prize was conditional—I had to pass a sort of test to make sure I was presentable, unlike the Nobel Prize, which you get even if you're a known racist or a misogynist!"* Two exquisitely polite Japanese men met with Daniel when he returned to Vancouver. "They asked a lot of questions and talked to my colleagues as well. We went out and ate a lot of good food, and it turned out fine in the end," Daniel concludes.

Their neighborhood survey a success, the Cosmos committee began preparing for Daniel and Sandra's weeklong visit to Japan. They requested photos and other documents for a booklet designed to tell the story of the candidate's illustrious career. The Paulys aren't big photographers, but Sandra managed to gather a few snapshots that told the story of her husband's journey from La Chaux-de-Fonds to Vancouver. The album opens with a black-and-white photo of Daniel, not unlike a Studio Harcourt movie-star portrait, but with lower light. A block of text in Japanese on the facing page recounts his heroic story. The Japanese text is accurate, but, thanks to a small translation error, the English version takes a more adventurous turn: "Dr. Pauly was then in effect kidnapped in his infancy by a family in the French-speaking part of Swaziland." Daniel finally got the African childhood he'd always dreamed of!

On a more serious note, the document also provides a summary of his scientific work, highlighting the fact that he had published over five hundred scientific articles and thirty books to date. As well, it contains a transcript of Daniel's speech from the award ceremony. He carefully avoids the subject of Japanese fisheries, but does insist in his introduction that, though there has been a lot of talk about pollution in the oceans (notably thanks to Rachel Carson and Jacques-Yves Cousteau), fisheries have not received the same attention. He then lays into those who still believe that climate change alone is responsible for the disappearance of terrestrial and now marine megafauna. Daniel goes on to discuss his work, mocking the French fisheries scientists who have

---

* Daniel is referring to James Dewey Watson, who won the Nobel Prize for Medicine in 1962 alongside Francis Crick for discovering the double helix structure of DNA—a discovery that was in fact based in large part on the work of a female scientist, Rosalind Franklin.

blamed "changes in species distribution" for the disappearance of fish stocks. Photos from the event show him, gray-haired and radiant, with the Cosmos medal around his neck on a rainbow-colored ribbon that reminds me of Bob Marley's signature scarf.

"We had a great week," Daniel confirms, "and I hope it helped, even just a little, to bring the problem of marine resource management to the attention of the public in Japan." Listing their activities, Daniel recalls, "I did three conferences, one in Tokyo, one in Kyoto, and one in Osaka. Then Sandra and I met the prince and his wife at the imperial palace. On the way out, the journalists didn't care about us. They were there to collect gossip about the royal couple, but of course we didn't say anything. Everywhere we went, we were treated like royalty. The one exception was at the Ministry of Fisheries, where I was received by some underling in a dusty office, and there was no exchange of gifts— it was their way of showing their disgust."

With the 400,000 dollars in prize money, Daniel finally paid off the couple's home—the upper floor of a townhouse with bottle-green trim set on a tidy, peaceful street on the UBC campus. Their next-door neighbors are none other than Carl Walters and Villy Christensen. With the leftover funds, Daniel created a scholarship to support students from developing countries who want to work in their country of origin. "I'd like to point out," Daniel insists, "that it's the only scholarship awarded by UBC that isn't named after its donor—it's also called the Cosmos Prize."

"Ah . . . everything was going right at UBC during that time," he sighs nostalgically. "I was the university president's favorite person." This state of grace made it possible for Villy Christensen, his longtime research partner, to gather funds for a marine ecology event focusing on Daniel Pauly. Besides, on May 2, 2006, the maestro would be celebrating his sixtieth birthday, and to mark the occasion—and because they'd nearly lost him a year earlier—Villy invited many of the people who had played an important role in Daniel's career. Each participant presented a piece of the scientific puzzle that constituted Pauly's global vision of the world's oceans, and their conference papers later went into a collective work edited by Villy and Jay.[7] Many of Daniel's old colleagues answered the call, and he was moved to tears by the sight of his former mentor, Gotthilf Hempel, accompanied by Cornelia Nauen, his Filipino friends Annadel Cabanban and José Ingles, and many, many more, all of whom had come to Vancouver to be part of his birthday celebration. From a more scientific standpoint, the event allowed his team to highlight the importance of "ecosystem approaches to fisheries." Behind the dry title is a very simple concept: in order for exploited fish stocks to be healthy, the ecosystems they live in need to be healthy, too.

RECONSTRUCTIONS

T
HE SEA AROUND US had taken on planetary proportions. Daniel and his team had proven themselves with their work on North Atlantic marine ecosystems and by exposing the Chinese administration's faulty statistics. The Pew Charitable Trusts rewarded them with funding that would make it possible to expand their analysis to cover all of Earth's oceans. Under Villy Christensen's supervision, they generated Ecopath-type models for all sixty-four of the world's marine regions. But, in accordance with the ecosystem approach to fisheries, they didn't limit their analyses to exploited fish species and their environments: they also included marine megafauna, the large animals that inhabit aquatic environments.

In the late 1990s, Daniel and his UBC colleague Andrew Trites began tracking down studies on marine mammals in order to accomplish just that. These creatures, after being harpooned to near extinction, are still regularly blamed for competing with fisheries. They recruited the German researcher Kristin Kaschner, with whom Daniel enjoyed many an intellectual sparring session in his second language while she worked on mapping the global distribution of 115 (of the 120 known) species of marine mammals and calculating their resource consumption.[1] The results: all the whales, dolphins, and seals in the world do indeed consume more seafood than all the humans, but the differences between human and animal diets are such that they rarely compete for resources, and when they do, it is mainly in the polar regions, especially the northern hemisphere. Those regions in which marine mammals and fisheries coexist only provide about 15 percent of the world's catch, however, and just 1 percent of the animals' diet. These figures are vitally important when it comes to confronting fisheries managers in certain countries, who have suggested simply killing off the remaining populations of marine mammals in order to "improve fishery yields."[2]

In the same vein, Daniel and Reg Watson codirected the PhD thesis of a young researcher named Vasiliki Karpouzi, who left her native Greece for cloudy Vancouver to calculate the nutritional requirements of the world's seabirds. After an absolutely tremendous amount of bibliographic research, Vasiliki was able to show that, in all, the 350 or so known species of marine birds consume a slightly smaller proportion of the ocean's resources than do fisheries. And like marine mammals, they mostly feed on organisms with no commercial value. However, she did identify a few areas where competition between seabirds and fishers is strong, namely the North Atlantic and the Mediterranean.[3]

BUT THE FISHING statistics on which these comparisons depended were false, and Daniel knew it. The data sent to the FAO by UN member states included, at best, catch volumes from industrial fishing, but never the total volume fished, which he knew had to be much higher because of bycatch, undeclared catches, and artisanal and sport fishing. So, Daniel threw his team into their biggest project yet: reconstructing those statistics to get figures that more closely resembled reality.

The process is incredibly simple, explains Dirk Zeller, who coordinated data reconstruction for 273 countries and overseas territories. First, they gathered FAO and other "official" statistics from 1950 to 2010, then examined them for potential problems. From there, the Sea Around Us team, with its dozen PhD students each paired with at least one local expert, began gathering nonofficial data to fill in the gaps. "The key part of the methodology proposed here is psychological," Daniel wrote. "One must overcome the notion that 'no information is available' . . . Rather, one must realize that fisheries are social activities, bound to throw shadows onto the societies in which they are conducted."[4]

Such shadows were precisely what the Sea Around Us team hoped to find by studying publications, technical reports, port authority records, and so on, in a multitude of local languages—sources that international researchers had generally failed to take into account. The Sea Around Us team could read in thirteen languages, including Mandarin, Japanese, and Russian, and Daniel recruited research assistants who understood the required languages to deal with the rest. "The rarer the language, the more Daniel enjoys decoding the text," they tell me in the hallway at UBC. The Vancouver-based researchers also drew heavily on British and French colonial archives and hunted down all the information they could find on the internet: newspaper articles, satellite images of fishing installations, even photos of trophy catches. This allowed them to go beyond the official statistics and reestimate the number of active fishing boats, their catch volume per species and per day at sea, and, ultimately, the total catch volume year by year for each country. This incredibly time-consuming, repetitive, and onerous work was carried out for every country and territory on the planet that had any coastline whatsoever. Even the Gaza Strip was included. The 325 local experts made enormous contributions, some of them over several years. Daniel's reputation helped generate interest in the project: some candidates had already crossed his path during his thirty years of world travel, while others jumped at the chance to work with the famous Professor Pauly.

The result: a huge chart covering all the fisheries in the world from 1950 to the present, a historic and geographic picture of unprecedented proportions. Their data was not as precise as the biomass per species figures painstakingly tabulated by countries in the Global North. But, as Daniel points out, "We gain nothing from the notion that only a select group has the key to understanding fisheries, especially if that key cannot open any doors outside a small number of developed countries."[5] "Yet," he continues, "it is only by making bold assumptions that we can obtain the historic catches needed for comparisons with recent catch estimates and thus infer major trends in fisheries."[6]

And these comparisons are informative indeed: although the FAO reported that catches had gone up over the decades to reach 86 million metric tons per year in 1996 and supposedly leveled off or declined only slightly from there, Pauly and his team showed that the "reconstructed catches" were actually one and a half times that. Additionally, though the "reconstructed catch" also reached its climax in 1996 at 130 million metric tons, it has been declining ever since then by a million metric tons per year on average. Most importantly, this drop is not the result of improved fishing regulations but of generalized resource depletion, with industrial fishers being responsible for three-quarters of the global catch. The authors also underline the fact that 10 percent of seafood caught each year is discarded, particularly as a result of shrimping, which is one of the worst offenders. Ten million metric tons of seafood, almost all of it perfectly edible, is pulled out of the oceans each year and wasted. In conclusion, the authors call on the United Nations to reinforce the agencies responsible for tracking fishing activities, adding that incomplete statistics also plague other domains. For example, although the FAO announced that deforestation had fallen by half in the first years after 2000 based on the reports of member states, satellite analysis showed that it had actually doubled.

The results were promising, but, once again, the work took its toll. When I saw him in Vancouver in 2015, Daniel and his coeditor Dirk Zeller were finishing their five-hundred-page *Global Atlas of Marine Fisheries* and an accompanying article that summarized the work.[7] "We worked like crazy—I'm scared to death," Daniel confessed, and he suddenly seemed older, worn out. He would often arrive at the lab unshaven and bent over by a terrible backache. Daniel knew that this would be his last big project, the result of a decade of hard labor—and also that he would probably be attacked and questioned mercilessly about the reconstructed catch statistics, which the leading lights of fisheries science would dismiss as far too esoteric.

But the study, published in spring of 2016, was a home run, and when I saw him again, Daniel looked ten years younger. Boris Worm and many others came out to support him, the media overflowed with praise, and their article was cited one hundred times in the first year after its publication. The following year, Dirk and Daniel won the Blue Marine Foundation's Ocean Award for their work.

AFRICA FOREVER

WEST AFRICA IS a region plagued by all of humanity's ills. The slave trade, colonial empires, civil war, desertification, pandemics, corruption—the nations of the Global North prefer to ignore the existence of economic refugees, but their methodical pillaging of the African continent goes on. Fisheries are no exception.

Since his first research mission in Ghana in 1971, Daniel has returned to Africa many times, learning about its geopolitics, taking a passionate interest in its decolonization, and shaking his head at the region's vagaries. He has also met the leading lights of French fisheries science there, particularly in Dakar. Indeed, for over fifty years, Senegal has hosted a rich community of French researchers whose work centers entirely on fish and fisheries. It all began in 1938 with the arrival of Théodore Monod, who directed the Institut français d'Afrique noire until 1963. Though Monod is best known for his expeditions in the Sahara Desert, he began his career as a marine biologist interested in fish. Following in his footsteps and those of oceanographer Anita Conti, several generations of French researchers have flocked to West Africa to study its fisheries. In 1950, the French development agency ORSTOM was established there, housing more than thirty French expatriates by the late seventies. Even today, Dakar is home to 115 French and Senegalese researchers, most of whom work for ORSTOM's successor, the IRD. Thanks to this European connection, Senegalese fisheries scientists often complete their studies in France—at the University of Western Brittany in Brest for the older generations or at the National School of Agronomy (ENSA) in Rennes for the younger ones—where they are coached in the productivist* doctrine of French fisheries science.

Since Ghana and his almost-hiring by ORSTOM in 1974, Daniel's relations with French scientists stationed overseas had been rocky—and they were about to get worse. West Africa, and Senegal in particular, was soon to become an important front in the war between Pauly and his detractors.

The Canary Current system, which sweeps down Africa's Atlantic coast from Morocco to Guinea, is one of the most fertile in the world. Historically, an incredible variety of marine life flourished there: pods of whales, sharks, giant rays, sea turtles, and swarms of seabirds. White clouds of such birds would guide local fishermen toward schools of fish so dense that they darkened the water. At dusk, lost sailors could follow flocks of terns and tropicbirds back to land. People who lived on the coast gathered various organisms, but especially grouper and sardinella. Despite their diminutive suffix, sardinella (which measure around eight inches in length) are in fact the larger cousins of the sardines commonly fished off the Atlantic coast of Europe. They are fatty and nourishing—though also full of bones. In the Wolof language of Senegal, they are called *yaboï*, and they feed an entire nation turned toward the sea. The Senegalese consume over 66 pounds of fish per person per year, covering three-quarters of their animal-protein requirements. Fishermen in pirogues† catch sardinella in purse seine nets‡ within thirty miles of the coast. The fish are then deposited on the beach where the women take over, drying, braising, or boiling them before preserving them in salt. The fish's strong odor permeates homes throughout the country, especially the poorer ones. Senegalese fishers also catch the "white grouper" (*thiof* in Wolof), a plump and subtly marbled fish that is an object of intense fascination for the

---

* The *Oxford English Dictionary* defines productivism as "the doctrine or theory that increasing productivity is the primary goal of socio-economic activity." (TN)

† A type of small, traditional banana-shaped boat usually made of wood and, in Senegal, often painted in bright colors. (TN)

‡ A large, pocket-shaped net that is closed around a school of fish.

Senegalese. Whereas sardinella live in the water column alongside other pelagic fish, white groupers prefer the seabed, where they can stuff themselves with crustaceans and smaller fish. They are caught using lines, drift nets, or bottom trawling. Unlike the common *yaboï*, *thiof* is a luxury good that only the well-heeled can savor in their *thiéboudiène*.*

"Foreign boats come and take everything." This is a complaint I heard constantly throughout the 1980s. At that time, my sister Isabelle was living with Serigne, who is from Ngor (near Dakar), and my family became part African. Back in the suburbs of Paris, I'd heard stories about a fishing village with narrow streets that resound with the call to prayer each day at sundown, a beach where beautifully painted pirogues come ashore, an enchanted island where locals sometimes sleep under the stars. After a childhood spent with my nose in plates of beef bourguignon and potato gratin, I ate my first mango and many other African delicacies, including *thiof*. Over dinner, I was told about the heroic daily lives of the piroguiers and the scrappy resourcefulness of hungry children who snatch small fish off the boats as they come ashore. My African family also complained about the armada of foreign ships that were pillaging African waters. This shameful practice had been going on for some time already and took several forms, but in essence, the whole world was helping themselves to the marine resources off the Senegalese coast.

Historically, tuna and sardine fishermen from Brittany or the Basque region were the first to set their nets off the coast of West Africa, beginning in the 1950s. Unsurprisingly, this was the precise period when a colony of French fisheries scientists set up shop in Dakar. Indeed, I have been told that the catch statistics for tuna are particularly well documented, and many a French scientist has made their career by analyzing them. These fishing activities, in which the Spanish also participated actively, continued over the decades in order to supply Europe with Atlantic tuna. In the 1960s and '70s, the Soviets arrived

---

* The national dish of Senegal, composed of rice, vegetables, and fish.

on the scene with their factory ships, mainly going after small pelagic fish off the coast of Mauritania and Namibia. Then, in the 1980s, boats from Asia, and particularly China, showed up. Even today, Europeans, Russians, and Asians are still putting a lot of pressure on West African fisheries. Their catches are partly legal, since tuna is mainly fished in international waters, and also because fishing agreements regulate the activities of foreign boats in the exclusive economic zones of some African countries: once an entry fee, sometimes substantial, is paid, the sea turns into an all-you-can-fish buffet for foreign vessels. These are referred to as "uncontrolled fisheries." As if that weren't enough, some boats operate completely outside of the law, a practice called "pirate fishing." One way or another, all these catches find their way back to Europe or Asia. Perhaps most scandalously of all, an increasingly large share of smaller pelagic fish caught using purse seine nets aren't even sold directly to consumers. Instead, they are transformed into fish meal and used to feed chickens, pigs, and farm-raised salmon. The rationale behind this process is dubious: it takes four or five tons of fresh fish to produce a single ton of fishmeal, and a lot more than one ton of meal to produce a single ton of salmon. Initially, meal factories were set up to process scraps, but for some time now, purse seiners have been fishing specifically to supply the factories, a setup somewhat obliquely referred to as "reduction fisheries."

I wouldn't make it to the fisheries center in Dakar until thirty years after Daniel, but finally, in October 2016, there I was, at the invitation of BirdLife International, a British NGO that was rightfully worried about the impact of fisheries on seabirds. I presented a synthesis of my team's work showing that shearwaters and northern gannets, though protected on their nesting grounds in Brittany and on the islands off the coast of Marseille, were not so carefully looked after during their migrations in and around West Africa. There, they fall prey to accidental capture by industrial fishing boats, which are in direct competition with the birds for sardinella. Even worse, we strongly suspect Chinese vessels of catching the birds on purpose. This totally illegal type of hunting supplies the Chinese market with much prized "wild flavors,"* the most sought-after of which come from endangered species. Several containers of frozen gannets and shearwaters have been seized in the last few years in Senegal and Mauritania—a nightmare vision for a marine ornithologist such as myself. Seabirds are the most endangered of all bird groups, and they are all too easy to capture at sea. "If a market for them develops in China, you can say goodbye to the last albatrosses and other petrels endemic to the tropical and Antarctic seas," warns Ross Wanless of BirdLife International. In Dakar, the work group in which I participated included both ornithologists and fisheries biologists, with many representatives from the Oceanographic Research Center of Dakar-Thiaroye (CRODT).

One of CRODT's executives, Djiga Thiao, presented a very enlightening overview of the state of fisheries in West Africa. According to him, "all the available research confirms that sardinella are overexploited," and he highlighted the fact that this "strong decline in fish stocks" is being caused by "all the maritime actors, including artisanal fishing." Additionally, "the *thiof* stock has been in a state of collapse for years, to such an extent that it has even been added to the red list of endangered species." I asked him about reduction fisheries, and he agreed that their "fast growth is very worrisome." A Mauritanian colleague stepped in to add that in his country, no fewer than twenty-seven fish meal plants are currently in operation, mainly to process sardinella brought in on Senegalese pirogues.

Thiao also mentioned the damage caused by illegal and uncontrolled fishing, as well as flaws in the surveillance system. During the Q & A that followed, the audience asked him to specifically identify the guilty parties. He hemmed and hawed some, but admitted that they were mostly Asian and European. A Catalan researcher pressed him for more information: "Can you give us an order of magnitude for this illegal fishing? Is it 10 percent, 50 percent, 100 percent as much

* Called *yewei* in Mandarin.

as officially declared fishing?" Djiga Thiao refused to answer: "I work in a fisheries institute—there are some questions I don't want to get involved in. There are some NGOs who give numbers, but I can only work with official figures."

And that leads us straight to the heart of the problem: What is the actual scope of overfishing in West Africa, be it from official or pirated catches? This question had preoccupied Daniel for a long time.

Since the early nineties, he had been reflecting on the case of Sierra Leone, where artisanal fishing could feed the local population much more sustainably than blood diamonds, which mainly enrich South Africans and their mercenaries. With support from the European Union, ICLARM sent the indefatigable Michael Vakily to Freetown to coordinate a summary report on local fisheries using the protocol already established by Pauly and company. The main objective was to rescue a huge amount of data stored in Russian on floppy disks from simply rotting away in the tropical climate. With help from local fisheries scientists trained in Moscow, Michael decrypted the archives and created a database. He also organized a national conference on fisheries in November of 1991 with help from the FAO, which took place despite an unstable political situation and made it possible to analyze three decades of data from the postcolonial period, during which aquatic resources were still abundant and anything seemed possible.

That same year, though, Sierra Leone collapsed into chaos— Vakily stayed as long as he could, but he was ultimately evacuated with his family in 1994, without finishing the conference proceedings. The civil war raged on until 2002, killing several thousand people and creating a whole generation of child soldiers. Rebels took Freetown and transformed the library of the fisheries research center that Vakily had restored into a kitchen, burning most of the historical data.

Twenty years later, Katy Seto, the PhD student Daniel had charged with reconstructing statistics for Sierra Leone, contacted Vakily, who, after traveling around the world to help construct FishBase, was back in West Africa. Always meticulous, Michael had carefully conserved the conference proceedings in electronic format. They were finally published in 2012 with a less-than-optimistic preface by Rashid Sumaila, who noted: "What Africans, and Sierra Leoneans in particular, had not counted on is that independence did not guarantee access to these natural resources—indeed the competition for those resources continues still, and as far as fisheries resources are concerned, it is largely the EU, and increasingly China, which wins, and not Sierra Leone."[1] Unfortunately, as we shall see, the same can be said for all of West Africa.

IN THE LATE nineties, Daniel renewed ties with Birane Samb, whose career he had been following and supporting from afar since their first meetings at different ELEFAN training sessions between 1984 and 1986. Since then, Samb had been to the United States and France, where he studied at the École centrale de Lyon to become a hydroacoustics specialist. In fact, the best way to detect schools of small pelagic fish is by using sonar, a method Birane mastered perfectly. Based at the CRODT in Dakar, he participated in numerous surveys of small fish off the coast of West Africa. Oddly enough, these campaigns were organized by the Norwegians using the *Dr. Fridtjof Nansen*—a huge, white-sided oceanographic vessel 187 feet long that sailed the world surveying fish stocks in developing countries for more than two straight decades without even once returning to its port of origin in Bergen.

As a regional coordinator, Birane had access to the *Nansen's* data on sardinella distribution. He noticed that stocks fluctuated greatly on the northern (Morocco) and southern (Senegambia) limits, but much less in the center of their distribution zone (Mauritania). According to the season, then, the entire sardinella population oscillated up and down a thin strip of coastal water, like liquid swishing back and forth in a tub. Birane concluded that it was essential to consider the sardinella stock as a whole, and that adopting a country-by-country management scheme would be a mistake. He mentioned this to Daniel, who encouraged him to publish his ideas and helped him write a note on the subject. The duo published an opinion piece alongside maps illustrating the sardinella's distribution zone that largely spoke for themselves.[2] They also took advantage of the opportunity to dispel several commonly held beliefs on fish stock management. In fact, in a famous article published in 1992, Baumgartner and a team of colleagues in California had suggested that populations of small pelagic fish fluctuate naturally, even in the absence of fisheries.[3] They reconstructed historical population numbers by analyzing layers of sediment on the ocean floor off southern California, evaluating the density of fish scales in each layer and using that figure to measure how plentiful fish had been over the last two thousand years. They concluded that during that time, the total biomass of small pelagic fish varied on a scale of magnitude of one to seventeen.

But, as Samb and Pauly pointed out in a paper in 2000, focusing on this kind of natural variation only incites fisheries managers to avoid taking action, no matter what: even if fish stocks collapse after being overexploited by industrial fishing, it's probably just due to natural causes, and the only thing to do is wait for the stocks to regenerate on their own. This is the argument used everywhere small pelagic fish are exploited, including in the most productive waters on the planet: West and South Africa, Peru, and California and Mexico. Samb and Pauly's counterargument, however, was based on spatial variability: as their analysis of the sardinella shows, it is dangerous to try to understand the qualities of an entire fish stock based on an analysis of a single part of its distribution zone. Almost two decades later, though, their warning still falls on deaf ears. As André Fontana and Alassane Samba point out in their excellent book on fishing in Senegal, "none of the efforts to coordinate fish stock management have come to anything."[4]

Samb and Pauly's article received only a moderate amount of attention from the scientific community, but it sparked a big fight. Gabriella Bianchi of the Norwegian fisheries institute accused the duo of publishing the data without her permission. Indeed, although the *Nansen* collaborated with coastal African countries, in practice local researchers never used the data they collected. As Daniel put it, "To use the data, they have to have permission from ministries in every one of the surrounding countries—there's so much bureaucracy that the data is, in reality, totally inaccessible. The *Nansen* program was regarded favorably, but there was a total lack of knowledge transfer, and that was its major flaw—only the Norwegians could use the data."

Birane Samb has been with the FAO since 2010, where he still works alongside Gabriella Bianchi. When I ask him about the problem of data sharing, his answer is diplomatic: "Usually, there is a data utilization policy in place, but it wasn't working at the time. It generated a lot of conflict. But ultimately, you have to look at the results—there needs to be a regional management scheme for sardinella; those stocks are the most widely shared in West Africa." But at the time of the article's publication in 2000, Gabriella Bianchi lost her composure. Though she knew Daniel already—he had been one of the readers for her doctoral thesis—the emails she sent after the publication of the editorial were, Daniel says, "a mixture of racism and vitriol" in which she accused him of "using and manipulating Birane Samb, because it couldn't have been his idea to publish that article." "I was so disgusted," recalls Daniel. "I never worked with her again." Their heated altercation proved productive, however—since 2006, the *Nansen* project has strengthened its partnerships with scientists in developing countries.

At the dawn of the new millennium, Daniel pressed on, working to create a large-scale synthesis of research on West African fisheries. "There were a lot of short-term studies," Daniel points out, "but it's only when you consider the long-term view that the worst effects of overfishing become visible." The project, called the Système d'Information et d'Analyses des Pêches (Fishing Information and Analysis System), or SIAP, was financed by the European Union. For once, Daniel accepted that he would not have the time to manage yet another project and handed off responsibility to Michael Vakily and Pierre Chavance—although Palomares and Pauly did write one lengthy report, the Sea Around Us's contribution to SIAP.[5] The project offered local partners the opportunity to publish ecosystem models for all of West Africa's maritime countries, which Villy Christensen then stitched together, demonstrating that the biomass of exploited marine resources in the region had fallen by 90 percent since 1960. In his editorial, Daniel points out that there has been "a massive decline of the abundance of larger fishes along the West African coast," news that he insists "should impact on negotiations about access by Distant Water fleets." This would indeed be the case, especially for Senegal, which has since made repeated attempts to rid itself of such maritime bloodsuckers. But, as the French proverb goes, when you chase the devil out the front door, he comes back through the window. In the last few years, foreign ships have continued their activities by adopting Senegalese nationality. In the Mediterranean, French fishers used this same tactic to continue chasing bluefin tuna. Until this loophole was closed by the collapse of the Gaddafi regime, French tuna boats simply became Libyan, filling the port of Sète with huge vessels sporting Arab-sounding names.

"THE WHOLE WORLD is pillaging Africa's marine resources. The Senegalese may have chased off the foreign fleets, but their own fishers are still seriously overexploiting their resources," explains Didier Gascuel. I contact Didier, an excellent interview subject, because of his contributions to SIAP and more importantly because, with Philippe Cury, he is one of the only French fisheries biologists of his generation to talk seriously about overfishing. As the head of the Association française d'halieutique (French Association for Fisheries Sciences), his words carry a certain weight, and he promotes the ecosystem approach to fisheries management with the EU in Brussels. Didier's history is an interesting one: he joined Pauly's school despite his early training in the "agronomics" approach to resource management. In France in the 1970s, students like Didier with a penchant for mathematics who were interested in nature and the ocean didn't waste their time with traditional university programs in ecology, which were underdeveloped at the time. Instead, Didier completed a series of preparatory courses after high school, then joined the highly prestigious École nationale supérieure agronomique (ENSA) in Montpellier in 1978 and subsequently landed a permanent position with the Institut national de la recherche agronomique (INRA) in Rennes. Agronomy departments in France trained generations of ecologists, profoundly influencing their philosophies. This training tended to discourage scientists from getting involved in public debate, something that is still difficult for French ecologists.

"Those programs were all about postwar productivism," Didier recalls. "When I think back on it, I turn red with shame, but at the beginning of my career, all I cared about was dead fish, totally divorced from their ecosystem. Everyone's ideas evolved over time, but we were all exposed to that way of thinking in the beginning. Daniel was one of the pioneers, a visionary who tried to reconcile exploitation and conservation. Even his critics have to admit that now. Our discipline was confronted with the problem of overexploitation early on, but we thought science could fix everything and we put our faith in it blindly, thinking fisheries science would necessarily lead to sustainable fishing. The trouble, though, is that everyone was in his own little world, working on a single fish stock and thinking that, even though things weren't going great, it would all work out in the end."

"We also lived in a time," Didier continues, "when scientists didn't talk to the public—nothing was more humiliating for a scientist than seeing his name in a big newspaper. Your colleagues would actually look down on you if you spoke to the press. This was still the case inside the French research community until the early 2000s. In North America, of course, it's the other way around—the best thing that can happen to a scientist is to get a big article in the *Washington Post* about his latest publication in *Nature*. The one in the newspaper is almost more important than the one in the scientific journal!"

Didier also visited Dakar in the early 1990s because, he says, "it was *the* center of excellence for studying fish population dynamics in the French-speaking world," and he has maintained strong ties to West Africa. Today, he is still a member of the scientific committee that monitors fishing agreements between Mauritania and the European Union. It was because of his expertise in this area that he joined SIAP, where he encouraged West African countries to pool their data. "Normally, they tended to keep their data to themselves," Didier notes. "I worked on setting up a data-sharing policy that would guarantee open access after a period of five years, and I'm quite proud of the result."

But Daniel's head nearly exploded: "At the time, it made me very angry—we had gotten funding to collect and organize that data, including historical information that we had thought was lost, and then it wasn't going to be immediately accessible to everyone. Data has to be accessible, because by working with it, researchers are indirectly helping the countries in question." Twice, once in the Philippines and now in Mauritania, Daniel had seen huge datasets, financed by tax dollars, "locked away by administrators and their passwords." In the big-data revolution that was beginning, Daniel took the side of "immediate open access." Didier doesn't agree: "If you tell the Africans that, they'll never put any of their data in the pool." After arguing over the problem several times, the two colleagues ended up in "a violent shouting match in the streets of Dakar after a long day of meetings." Each called the other an "idiot." "Daniel was very terse, very authoritarian back then," Didier recalls. "He wasn't used to people standing up to him, but in a way, I think he enjoyed a good argument."

Indeed, several years later, in 2004, when Didier visited Vancouver for the World Fisheries Congress, he fell in love with British Columbia and immediately sent Daniel an email asking if he could spend some time at UBC. "He got back to me right away with a big 'YES.'" Didier moved to Vancouver in 2006, as soon as he could find the funding, and in the meantime, Daniel had become a new man. "His stroke changed him a lot," Didier observes. "He pays a lot more attention to others." The two discovered that they had something more in common: a Marxist youth. "That helped us bond," comments Didier, whose whole career was transformed by his time in Vancouver. "I realized that research and activism can go together, that history is also written through individual political engagement." Didier used his trip to Canada to finalize the biggest conceptual advancement of his career: the EcoTroph model. Like Ecopath, Didier's new algorithm recreates marine ecosystems, but it sets aside the notion of species, using functional groups, the different "trades and professions" that structure every ecosystem. EcoTroph is now used to analyze the ecological impact of fisheries all over the planet.

Daniel and his team had already revealed how overzealous Chinese bureaucrats were padding catch levels in their domestic waters, causing an overoptimistic view of global catch trends. In its mad dash for the ocean's remaining fish, though, China quickly ran up against the same problem as Europe and Russia: its own coastal waters were becoming rapidly impoverished, and Chinese fisheries were forced to expand their horizons. At the dawn of the twenty-first century, armadas of factory ships and a multitude of trawlers, purse seiners, and long-liners, already present on the open ocean, set out to plunder the coastal waters of Asia, South America, and Africa—their crews working in conditions so appalling as to be described as modern slavery. Even when these activities were legal, specialists suspected that their catch volumes were underreported so as to minimize the appearance of overfishing. "In 2004, when SIAP was finished, I still didn't understand the magnitude of the problem," admits Daniel, "and then I happened to read the 2008 French edition of Beuret et al.'s book *Chinafrique* and the fog lifted."[6]

To fill in the gaps in the FAO's data, the Sea Around Us team estimated the actual Chinese catch using a host of detective methods that they later applied on a global scale.[7] Using over five hundred sources (scientific publications, reports, news articles, fishing firms' websites, and other online information) in several languages, including Chinese, they discovered that Chinese ships were fishing in the two-hundred-mile exclusive economic zones of ninety-three other maritime countries, and that they could be found everywhere except for the Arctic, North America, and the Caribbean.[8] They estimated that these 3,400-some ships capture 4.6 million metric tons of fish per year, a total market value of nine billion dollars, or approximately four times the figure reported by the Chinese authorities. Most disturbingly, Daniel and his team estimated that Chinese fisheries were pulling 3.1 million metric tons of fish out of African waters each year. Once published, the study attracted a lot of media attention, including in China. The

Pew Foundation released a series of maps entitled "Where Chinese vessels fish," with the biggest point on the map hovering over West Africa. "When it comes to Chinese fisheries," concludes Didier Gascuel, "they're almost always pillaging, so when Daniel's study came out, everyone agreed it was good to hit them where it hurts."

This information increased Senegal's displeasure with foreign fisheries, but the Asian pro-fishing lobby struck back quickly and in a totally unexpected way. "The closing conference for SIAP had been such a big success, mainly because it had attracted the attention of the political sphere in Senegal. So, it was hugely disappointing, just a few years later, to see fisheries departments in West African countries saying that whales were behind the decline in fish stocks!" Daniel explains regretfully.

Smoke came out of Daniel's years when he understood the reasons for this strange new attitude: Japan, Norway, and a few other countries had been campaigning for decades to continue whaling. Within the International Whaling Commission, their strategy was to rally as many developing countries to their cause as possible, influencing their votes by any means necessary. "The Japanese bribed officials from several African countries, including Senegal, to say that whales were eating the fish, which is absolutely impossible because whales go to West Africa to reproduce, and everybody knows that they don't eat during that time," Daniel points out. "It was depressing because several decades' worth of research had just proven that overfishing (by humans) was the real problem."

The WWF hastily assembled a work group in Dakar in May of 2008, of which Daniel was a part. The event attracted an extraordinary amount of attention from the media, opening with press conferences by several ministers. Happily, the scientific evidence was very much in the whales' favor, and coastal residents love the creatures—the African politicians quickly abandoned their noxious rhetoric. Daniel returned to Vancouver, with "a heavy heart," and published a scathing editorial in which he openly denounced the corruption of the Senegalese political elite.[9] The following year, he coauthored an article in *Science* with Leah

Gerber of Arizona State University, announcing that massacring West Africa's remaining whales would not create any benefits for fisheries, whereas even slightly reducing the pressure of industrial fishing on marine resources would lead to noticeable improvements.[10]

"West Africa is the black hole of illegal fishing, especially the northwest." Daniel knew it already. But the first reconstructions completed by the Sea Around Us, particularly the ones for Chinese fisheries, did not differentiate between legal and illegal catches. To clear things up, Daniel needed a strong ally, a bulldozer capable of smashing the status quo and breaking the policy of silence. Dyhia Belhabib fit the bill, a young woman born in Algeria who has the determination of an Antigone. Not wanting to live in her native country, nor in France like a lot of Algerian expatriates, she enjoys life in Vancouver, a "neutral territory." We ate lunch together a few times during my stay in British Columbia, and I quickly sensed the magnitude of her talent and fury.

After studying at the École nationale supérieure des sciences de la mer in Algiers, then at the University of Quebec at Rimouski, she "didn't think counting otoliths was very sexy" and discovered that Daniel Pauly, the man who had written so many papers, was still alive. She arrived in Vancouver in January of 2011 and joined the Sea Around Us as a volunteer to help update their database of the world's marine reserves. When Daniel spotted her, he asked, "Why are you volunteering?" "Because," Dyhia answered, "I want you to direct my thesis." Daniel soon hired Dyhia, and she started off quite logically by reconstructing fisheries statistics for Algeria. "I'm really interested in how fishing affects people. Sometimes, it's depressing, but that reconstruction was an essential first step for what I wanted to do next, which was analyze illegal fishing and evaluate the equity of fishing agreements." After that, Dyhia estimated the "official and unofficial" catch volumes for the fisheries of twenty-two African countries with Atlantic coastlines, from Morocco to Namibia.

"I liked to start by studying the country's history, especially during the colonial period. Then, I would examine the FAO data and look for

shadowy areas, like periods where catch volumes stagnated in a suspicious way. Then, I did the bibliography, compiling sources in English,
French, Arabic, Portuguese, and Spanish, which allowed me to extract
data by fishing sector, then do the reconstruction. Of course, I had the
whole Sea Around Us team behind me, and when I got stuck, I could
always go see Daniel—he's like a living historical and fisheries dictionary." To estimate catch volumes for some areas, Dyhia was forced to
rely on a few nontraditional methods, including photos and videos of
sport fishing she found on the internet—mostly big white men showing off the plump fish they'd caught in African waters. She even used
Google satellite images to find fishing villages that may have been "forgotten" by the official statistics.

"Once I finished the reconstructions, I contacted our African partners so that they could check the information. Obviously, I didn't have
absolutely everything, but at some point, you have to know when to
stop." Dyhia confirmed how important artisanal fisheries are to the
survival of the West African coast's poorest inhabitants, while also
demonstrating the fragility of their activity: artisanal fishermen are
expanding their efforts even as the resources on which they depend
grow scarcer. She studied Senegal closely, estimating that illegal fishing
represents an annual revenue loss of 300 million dollars for the country.
Lastly, she showed that European fisheries operating in West Africa
declare less than 30 percent of their actual catch (1.6 million metric
tons per year), and the Chinese less than 10 percent (of 2.3 million
metric tons per year). Even if the Europeans and the Chinese toss a few
coins to the African countries, they are still getting the better end of
the bargain: fishing agreements only cost about 8 percent of the total
value of the European catch, while the Chinese get an even sweeter
deal by constructing various projects whose value was estimated to
represent about 4 percent of their total catch value. The study clearly
indicated that European fishing boats were guilty of misconduct—
Spanish and French ships were increasingly failing to declare the true
volume of their catches.

Dyhia wrapped up her thesis in three years, a rare feat in North America, and defended in December of 2014, eight months pregnant. "I absolutely had to finish before the baby was born, so I really picked up the pace—it was exhausting," she admits. Unsurprisingly, her studies made a lot of noise, giving the African countries everything they needed to expel foreign fisheries. "We must close the door on the elephant of industrial fishing," Dyhia states. "In Senegal, the revolution has already started—fishing agreements with Europe are limited to tuna. Gambia has outlawed all industrial fishing, and Liberia is holding its own against the Koreans. For the other countries, though, it's more complicated. In Sierra Leone, illegal fishing increased during the Ebola epidemic. Mauritania put fishing quotas in place, but they still have a big surveillance problem. And things are not going well in Morocco, which bought drift nets from the Europeans after the EU banned them in 1992. The collateral damage caused by fishing is huge, and the Moroccan government just continues abusing the marine resources off Western Sahara. The problem is that all of the development projects around artisanal fishing in Africa have failed—they always lead to overcapacity and end up increasing poverty. It's important to allow local fishers to benefit from local resources, but the infrastructure on land has to be reinforced first, in order to increase the market value of the fish."

GOVERNMENTS AND NGOS alike applauded the Sea Around Us's work in West Africa. "It was a real wake-up call," insists Charlotte Karibuhoye of MAVA.* "Usually, you see this kind of study presented at international conferences where the audience is all from the same milieu, all scientists who already agree with what is being said, who want to test their theories. Pauly and his team reached other types of actors, decision makers, civil society. They got people thinking."

---

* MAVA: a nature foundation founded by Swiss ornithologist Luc Hoffmann that is very active in Africa.

"Monitoring of fishing activity was reinforced, and fines went way up," remembers Camille Manel, of the Directorate for Sea Fishing in Dakar. Indeed, around that time, the charismatic Haïdar el Ali rose to power. Initially director of a diving school, the Oceanium in Dakar, which later became a center for the protection of the oceans, he was named minister of the environment in 2012, then minister of fisheries in 2013. He launched a campaign to take back Senegal's national marine resources, going so far as to solicit Sea Shepherd* to police Senegalese waters. The hour of victory came early in 2014, when an illegal Russian trawler, the *Oleg Naydenov*, was seized by the Senegalese navy. The prize was exhibited proudly in the port of Dakar until the owner sent the government a check for 600,000 euros.

The case for fish seemed to be moving forward as well, at least in Senegal, but some scientists dug in their heels: a Franco-Senegalese team published an especially negative analysis of Dyhia Belhabib's work.[11] Basically, they accused Belhabib, Pauly, and their African partners of using erroneous data and largely overestimating catch volumes, particularly for Senegal. Indeed, when I first began looking into the project, Didier Gascuel had warned me: "You're going to hear a lot of bad things about the reconstructions, especially for Senegal."

I speak at length with Francis Laloë and Alassane Samba, who are retired from the IRD and the CRODT, respectively, and who both dedicated their careers to studying artisanal fisheries in Senegal. They were indeed fuming over Dyhia's work. Despite these tensions, though, Alassane Samba very generously invites me to his home in Dakar. "It hurts that she [Dyhia] didn't work with the people who had put the statistical systems in place—she didn't listen to anyone." He worries about the consequences of overestimating catch volumes in Senegal. "The Russians used those numbers to argue that Senegal has huge stores of resources we aren't using." He relates that Dyhia initially corresponded with local experts from the CRODT and IRD, but that

---

* Sea Shepherd: an NGO known for aggressively interfering with fishing vessels, especially whaling ships.

disagreements got in the way, which was unfortunate for everyone. Dyhia, for her part, blames "the CRODT and IRD's lack of transparency when it comes to their data," and remembers "slamming [her] fist on the table after being yelled at"—she ultimately published her paper against the advice of certain local experts.

Daniel savors these skirmishes with gusto—conflict with the old IRD researchers was to be expected, and Dyhia's spark reminds him of himself at that age. He helped her write a "political" response, alternating between an authoritative tone and a derisive one designed to take their detractors down a peg.[12] "I didn't need all that to know that the whole world is pillaging Africa's fishing resources," comments Didier Gascuel with an air of amusement. "But controversy is the norm with Daniel Pauly. You drop him into a community that's running smoothly, and the next thing you know, it explodes. A lot of people don't like that—he just has this aura, and you either love him or you hate him."

Still, everyone can agree on the main issues: on a humid night in Dakar, I speak at length with Alassane Samba about pirate fishing, overcapacity in artisanal fisheries, the disturbing development of fish meal plants—"an aberration!" he calls them—and the frustration he feels as an ecologist. "Our politicians are useless—they only listen to scientists if it's good for their bottom line," he explains. "The last few years, surveillance has been basically nonexistent. The proceeds from the sale of fishing rights aren't being reinvested in enforcement. Currently, there's a single aerial patrol per month, with a French plane, and everyone knows about it beforehand. The state has a real problem maintaining sovereignty over its fishing resources."

Djiga Thiao, who also coauthored the response to Belhabib's article, minimizes the conflict with the Sea Around Us team: "We tried to avoid settling scores... It's true that the official fishing statistics aren't trustworthy, everyone agrees about that, even in the 'rich countries'... Daniel is very good at making pleas, and he attracts a lot of attention. He's not afraid to take things a step further, go beyond the usual limits... He has an audience here—he engages in provocation, and people are too scared to contradict him. He attracts a lot of media attention, too, and we admire him for that. When he comes to Dakar, it's in all the papers the next day. The NGOS pick up what's been said, sometimes exaggerating it a little, but it gets people's attention . . . Despite all the criticism, he might actually be having a positive impact on the management of West African fisheries."

DANIEL HAS BECOME an authority figure for several generations of African ecologists. None of them expected to meet a half-Black man. "My African students were in awe," recalls Coleen Moloney of the University of Cape Town in South Africa, who invited Daniel to the first post-apartheid conference on marine ecosystems in 1996. Gotthilf Hempel laughs heartily as he recounts his own experiences in Africa, where mentioning that he was Daniel Pauly's thesis director causes him to rise instantly in the esteem of his African colleagues. "Daniel is worth ten of you or me," asserts Philippe Cury of the French IRD. "Just by himself, he's managed to change things more than all our development projects combined—and most importantly, he's always available to help students from the Global South. He spends a huge amount of time and energy on them—if some guy from some out-of-the-way place asks him for help, he jumps on a plane without thinking twice. In Ghana, I saw him leave an international conference because a local student asked him for advice. They got in a taxi and drove fifty kilometers to see the site he was studying, and three months later, they published a paper together." For his part, Daniel is a little more circumspect: "Most of the African researchers I've trained fell in line and went to work for the establishment—in the end, nothing changed."

Before leaving Dakar, I stop to see Serigne, who treats me to *yassa** on the roof terrace of his house in Ngor, which has a fantastic view of the Atlantic, where we can see local kids surfing. After lunch, I take a dip in the waves and we relax on the beach, sipping soda. Not far from where we are sitting, I notice a few Swedish fiftysomethings drinking

---

* Chicken with onions.

beer, surrounded by young—supposedly legal-aged—Senegalese girls in heavy makeup. Africa is selling its youth, too. I knew it already, but it always hurts to watch.

FRENCH ALLIES

A CONSTANT PRESENCE AT meetings of the French fisheries science community since the turn of the twenty-first century, Philippe Cury's mustache and I have crossed paths in Paris, South Africa, California, and numerous other places around the world. And though some claim it's gone out of fashion, I think it communicates the elegance of a careful strategist, nicely complementing its owner's observant gaze.

Along with Jacques Moreau and Didier Gascuel, Philippe is one of Daniel's few scientific and political allies in France, where the research community in marine sciences has long been suspicious of his work, or rejected him entirely. A landlubber from Montargis, south of Paris, Philippe Cury studied agronomy in Rennes before going to Dakar for a year and a half of civilian service in 1980. There, he worked on artisanal fisheries before returning to France to complete his education in biomathematics in Paris and join ORSTOM. What followed was a brilliant career, mainly abroad, where he spent thirteen years in Africa and five in California before his current post in Brussels, where he defends the IRD's interests before the European Union. Between all his traveling, he directed the Centre de recherche halieutique (fisheries research center) in the French port city of Sète and made his mark on the discipline with a number of important publications. In particular, he became known as *the* champion of the ecosystem approach to fisheries management in France. I personally admire Philippe because in 2008, at a time when overfishing was a taboo subject in France, he published a book on the subject, entitled *Une mer sans poissons* (A sea without fish) with Yves Miserey, a journalist at the conservative newspaper *Le Figaro* who could hardly be described as a bleeding-heart leftist. The authors recount the sad history of ocean fisheries in France and elsewhere, leaning heavily on the works of Pauly, Worm, Jackson, and many others. Their conclusion is unforgiving: "In France, ocean conservation and the protection of marine species always comes after social harmony—a harmony so fragile that it can be broken by a couple of fishing boats blocking a harbor."[1] To back up their statements, the authors provide a plethora of examples. "Bit by bit, overfishing has destroyed everything—today, 95% of the fish caught in the Bay of Biscay are less than 23 centimeters (9 inches) long."[2]

Philippe and Daniel have known each other "forever" but can't recall the exact date of their first meeting. It was almost certainly at a research laboratory in Monterey, California—the same town where Steinbeck set his novel *Cannery Row*. But was it in the mid-eighties or the early nineties? In any case, on that fateful day, Daniel, intense as usual, had crossed swords with an American colleague. "Daniel wasn't afraid to go after people, especially in his younger days—he could be very verbally aggressive when he came in contact with people who were set in their ways intellectually," Philippe tells me when we meet in his comfortable Brussels office. "I've seen him stand up in the middle of a conference to harangue the speaker," he adds. In Monterey, Philippe was asked to calm things down, and he soon came to like this strange Frenchman from the Philippines who spoke with a Swiss accent. "When you spend time with Daniel, you realize very quickly that he can think ten times as fast as you," says Philippe. "He's more of a craftsman than an intellectual—there's a strategic aspect to his work where everything fits into everything else, like a set of Russian dolls. He always knows what he's doing. It's like climbing a huge cliff—you have to see it as a whole in order to find your footholds and climb up little by little, all while helping other people see it, too. The way he goes about things and helps others is very unique, and it's also part of the political aspect of his work."

"We became friends," Daniel says simply when I ask for his side of the story. Philippe became the proverbial president of Daniel's French fan club and the likely instigator behind many of the honors he has since received there. After being invited to the seminar in Vancouver where they celebrated the sixtieth birthday of the colossus of marine ecology in 2006, Philippe returned the favor, rolling out the red carpet for his friend at the fisheries research center in Sète, where Daniel spent four months in 2009. "But it was a fiasco," Philippe admits with a sigh. "Daniel needs to talk, and my colleagues snubbed him—I couldn't understand why... sometimes people are so petty..." I first met the Pauly-Cury duo at the Montpellier aquarium in spring of that year. Daniel was doing a presentation for the general public on Darwin's voyage around the world. He looked exhausted, hardly speaking when Philippe—in low spirits himself—introduced me to him after the event. Although Daniel likes to visit France—especially if he can make a side trip to La Creuse—he has never been terribly impressed with the way his native country handles environmental issues. In particular, he remembers the Grenelle de la Mer* in 2009, which he attended at Philippe's invitation: "I was there for two days. Of course, the ship-owners and the fishing lobby had the upper hand—they were the ones who were dictating what was good for the oceans. Can you imagine?" Daniel laughs out loud. "I started talking right away about the problem of subsidies that increase overfishing, and the whole room answered, 'What subsidies?'—and it's true, in France there are no subsidies for fishing, only 'adjustments.'"

As is often the case with the "consultative workshops" to which Daniel is invited, the rules had been set in advance with "specific terminology" being used to avoid any "slipups." "In France, IFREMER† are the main architects behind this language manipulation campaign," Daniel declares. "For example, they talk about 'maritime

---

* A high-profile forum on the future of marine environments hosted by the French ministries in charge of sustainable development and the oceans.
† Institut français de recherche pour l'exploitation de la mer (French Research Institute for Exploitation of the Sea).

professionals,'* which allows them to leave out researchers, who have often studied these questions for decades... Aren't we professionals, too?" Daniel chafed at these restraints: "I have the luxury of not obeying the rules because I'm not based in France and my university isn't going to put pressure on me." No one backed him up, though, and he was tempted to leave the conference after the first day. Daniel was already preparing to depart for Hamburg when Claire Nouvian of the environmental nonprofit BLOOM† managed to pull some strings late in the evening and get a meeting with the minister of the environment, Jean-Louis Borloo, at 2:00 PM the following day. "I got my audience," Daniel recounts. "Borloo arrived late, drunk, and immediately launched into a monologue about Obama, who he didn't think was so smart, talking about everything that Sarkozy was going to do better... It was unbelievable—because I'm Black, he made a nonsensical connection between me and American politics. Eventually, we managed to talk to him for five minutes to ask for marine reserves, a reduction in fishing effort, et cetera, but it was all for nothing. It made me sick. We were in this incredibly beautiful palace with seventeenth-century frescoes, and this little man was supposed to represent the French Republic."

But things were already changing and a few people began to break the silence. In 2008, researcher Benoît Mesnil retired from IFREMER in Nantes and published a scientific article titled "Public-Aided Crises in the French Fishing Sector."[3] This dense eleven-page paper told the story of forty years of conflict that was only now coming to a head. Mesnil thanked his colleagues at IFREMER for their input, but insisted that his statements were his alone—and he hit hard: "The French fishing industry has been drowned by subsidies meant to keep it afloat." He explains that the fisheries present themselves as victims of "the system" in the course of violent conflicts, asking for more and more money, which they always get. To support his argument, Mesnil cites

---

* *"Professionnels de la mer."* (TN)
† See bloomassociation.org.

the exact amount of public money involved: over 800 million euros in 2006, while the commercial value of all the seafood disembarked in French ports was barely one billion euros. "Despite heavy investments in fleet capacity, the French [fishery] trade deficit has consistently become worse in mass and value—and likewise for most EU countries," Mesnil writes. His analysis also includes a sociological perspective: "In reviewing hundreds of pages of French fishing magazines I have never encountered the word 'profit' in interviews or speeches of industry members: [it] looks like a taboo word, for administrations and politicians as well." According to Mesnil, this vicious cycle is facilitated by the state, which offers up huge subsidies without asking for anything in the way of "wiser use of marine resources, more attention to the market, improved social relations, provision of more accurate information on their activities, or stricter compliance with legal obligations in general." "In other words," explains Mesnil, "the subsidies were devised to simply continue 'business as usual,' with all its wrongs. Indeed, some subsidies have exacerbated the critical processes (open access, overcapacity, indebtedness, excessive effort, and/or non-selective fishing)." He then points a finger at French politicians: "None of those who had a role in the recital of crises reviewed above was ever admonished for the poor return to the nation of the subsidies he had decided. Most even had a brilliant career." And he doesn't leave fishermen unscathed, either: "The generation of now retiring skippers, who benefited from the generous aid policies, is among the most vociferous in demanding measures for the young fishers, and it is tempting to ask them what they did to make the profession attractive to their children and other youngsters." And to top it all off, Mesnil concludes, the fishing industry is still "terribly medieval," with rates of accidental deaths among the highest of any profession.

The French press, after years of silence, covered and even expanded the debate. Two profiles on Daniel Pauly, one in Le Monde and one in Libération,[4] as well as a radio interview on France Inter in spring of 2009, attest to this change of pace. Daniel's voice on the national radio is surprisingly soft: "We think the sea will save us, but it can't, not anymore," he says with the empathy of a doctor consoling a dying man. "We are experiencing a global biosphere crisis, confirmed by the recent financial crisis."* When the journalist asks which measure needs to be undertaken with the most urgency, Daniel responds, "In France? The application of the laws that are already in place and the application of the law of the market, which is being destroyed by subsidies." Very calm during his time on France Inter, Daniel let loose with Michel Henry of *Libération*, whose article began with the hook, "Careful with this guy, he'll scare you stiff!" "This tall, strapping man," Henry continues, "in good shape for a sixty-two-year-old, has a big advantage: he's a talker, more of a mockingbird than a monkfish." The article is particularly well researched, with a summary of Daniel's extraordinary life, from his early exploits to the causes he has taken up more recently. "It's easy to say that scientists are idiots because they don't go out and fish," Daniel explains in the interview. "It's as if politicians had the choice between taking advice from doctors or sorcerers, and they chose the sorcerers." Philippe Cury, who dipped into his address book to help organize this media crusade, adds his two cents: "Daniel is denouncing the fact that fishing is managed like an agricultural product, something we can control. In fact, we don't control anything."

And so, fisheries advance blindly, ever further, ever deeper. A study published by Telmo Morato of the University of the Azores in conjunction with the Sea Around Us showed that the average depth of trawling operations has increased substantially since 1950.[5] This meta-analysis, which covers every ocean on the globe, shows that "fishing down the deep" has intensified since 1980 to reach an average trawling depth of 200 meters (656 feet) in the North Atlantic and 600 meters (1,967 feet) in the Antarctic. Try to imagine what bottom trawling looks like: a huge pocket up to 115 feet tall (as high as an eight-story building) and 330 feet wide (about the length of an American football field). Pulled along by a boat with several thousand horsepower, this net is held open by two "otter boards," panels that can weigh over 2,000 pounds,

---

* The interview took place just six months after the 2008 financial crisis.

which are pushed outwards by resistance from the water. In the 1980s, researchers attached cameras to a bottom trawler for the first time in order to observe what was happening beneath the surface. I will never forget the shock that ran through me the first time I saw those videos of what looked like a giant bulldozer destroying the underwater landscape. The fish they tear from the deep are rattails, lings, and black scabbardfish. You've never heard of them because they're ugly little suckers—only their meat ends up on your plate. When left alone, they grow slowly, reproduce late, and live to be fairly old—meaning these species are among those most easily endangered by the blind, nonselective practice of bottom trawling.

Philippe Cury adds that "each year, trawlers cover half of the area of the world's continental shelf, which is twice as large as the United States of America, and five hundred times as big as the area that is deforested each year. Some of the most productive zones are trawled up to eight times a year in the North Sea, and between 25 and 131 times per year in certain estuaries."[6]

IN FRANCE, THE debate over bottom trawling turned out to be an intense one. I discussed these events with Claire Nouvian of the NGO BLOOM over dinner at her home in Paris in 2016. Daniel had just won the Albert I medal from the Paris Oceanographic Institute, and the ambiance was festive. Claire is a journalist who turned toward environmental causes after visiting the Monterey Bay Aquarium for a story in 2000. There, she discovered "the tragic destruction of ancient underwater ecosystems by a handful of industrial fishing vessels."[7] This indignation "turned into a lifelong crusade against super-powerful lobbies." Claire published *The Deep*, a richly illustrated volume that was eventually translated into twelve languages and that allowed her to create BLOOM.[8] In 2008, she spoke with French president Nicolas Sarkozy about the future of the oceans and collaborated with his government "during the first two years of his mandate, when a 'flurry of environmentalism' briefly took hold of the Élysée."[9] The Grenelle de la Mer resulted in the creation of a "deepwater fishing initiative."

Claire soon discovered that "industrial fishing was being protected at the highest levels of government, with the complicity of the president of IFREMER... After several twists and turns... the nonprofit organizations abandoned the initiative in July of 2010."[10]

The fight continued, however, at the European Union's headquarters in Brussels, where the fishing commissioner, Maria Damanaki of Greece, "demonstrated the courage and will to put an end to decades of destructive fishing." But negotiations soon bogged down "because of the opposition of a single commissioner: Michel Barnier, former French minister of agriculture and fisheries, who was recruited by the lobbies at the last minute."[11] His refusal was all the more surprising because only "between 44 and 112 crew members are employed aboard the ships that practice deep-water fishing, which corresponds to 0.2 to 0.5% of French crew active in 2012."[12] BLOOM didn't give up, though, and launched (among other initiatives) the most-signed petition in French history, which ultimately convinced IFREMER to revoke its scientific endorsement of the deep-ocean fishing lobby.

In 2013, French supermarket chains Carrefour, Casino, and Auchan announced that they would stop selling seafood from deep-ocean fishing operations, but Intermarché, the most implacable member of the group and the only one to own a fishing fleet, waited until 2016 to follow suit, promising to halt the sale of deepwater species by... 2025. In the end, after several reversals, and thanks in large part to the indefatigable Claire Nouvian and her army of activists, Ségolène Royal (minister of the environment under President François Hollande) championed regulations on deepwater fishing, which were finally passed by the European Union in June of 2016. "Our political systems are corrupt, sick, and under the influence of toxic industrial lobbies," Claire Nouvian concludes, "but with a big enough call to arms, it is possible to get things done, even despite the neoliberal wind that is blowing across Europe, where the reddest of red carpets are currently being rolled out for the most polluting and destructive companies."

DANIEL WOULD SOON work to determine the importance of this very mechanism—the impact of consumer activism and public engagement on the management of the oceans—with the help of yet another female colleague. According to a persistent urban legend, their first conversation went something like this:

> Daniel: "Give me one good reason to direct your thesis."
> Jennifer: "Because I'll make you famous. And at some point, you might say yes just so that I stop coming to your office to bother you about it."

Daniel hesitated because Jennifer Jacquet came from the United States, and he usually preferred to mentor young researchers from the Global South, often turning up his nose at graduates from the big American universities. Besides, Jennifer is more of a sociologist than an ecologist, and in 2004, Daniel was looking for students to reconstruct catch statistics from as many countries as possible, not to write pretty speeches. But Jennifer wouldn't give up—she had come too far and waited too long to fail so close to her goal. Originally from rural Ohio, she rallied to the environmentalist cause at the age of nine, and later worked for Sea Shepherd before discovering Daniel's work during her studies at Cornell.

"One of the many papers we had to read was 'Fishing down' . . . You can really detect something about someone's personality through their writing . . . just some of the ways he was phrasing things . . . the elegance in the first couple of paragraphs of his *Science* paper. I thought, 'I want to meet this person,'" she tells me when we meet at her office at New York University in downtown Manhattan. She is currently a professor of environmental studies there, and on this November morning in 2016, Jennifer does most of the talking. Besides being a professor, she is also a blogger, journalist, and essayist who speaks with the practiced ease of a professional communicator. Her website suggests contacting her agent to schedule any conferences. "Before Daniel's stroke, he was a completely different animal. It was so obvious that he was the best, and he won me over right away. I didn't even apply to any other PhD programs."

Daniel eventually gave in, but Jennifer had to wait until 2005 and the end of Daniel's brief convalescence before starting her thesis. Daniel asked her to reconstruct fishing statistics for Mozambique and Tanzania. She got to work, but also began developing "little projects on the side." In fact, Jennifer was mainly interested in the notion of individual and collective responsibility vis-à-vis the environmental crisis, and specifically in the strategies that make it possible for NGOs to twist the arms of huge multinationals with the public's help.[13] One such strategy is ecolabeling: the USDA Organic label in the United States guarantees that products come from organic farms, while the FSC label promotes responsible forest management. In the marine sector, the WWF partnered with Unilever, the biggest distributor of seafood in the world, to create the MSC (Marine Stewardship Council) label for sustainable fisheries in 1997. You can easily find the label—sporting a stylized fish on a sea-blue background—in your local supermarket, particularly on frozen Alaskan pollock. No fewer than 312 fisheries all over the world are currently allowed to use the label.

Initially, Daniel supported the MSC initiative of enhancing consumer accountability. "Eating bluefin tuna in a sushi restaurant is no less harmful than driving an extravagant car like a Hummer or harpooning a manatee," he wrote.[14] He gave generously of his time to help establish criteria for sustainable fisheries. But when Jennifer asked him about labeling in 2005, he replied, "Unfortunately, I don't think that's the right way to solve the problem [of overfishing]."

This remark troubled Jennifer, who had been convinced of the immense power held by consumers and the importance of ecolabeling, and she decided to take a closer look. She quickly went from surprise to disillusionment, and in 2007 the newly formed Jacquet/Pauly team published their first opinion piece[15] questioning the effectiveness of the MSC label: Asia consumes more than two-thirds of the world's seafood, and that market is totally immune to ecolabeling. The same thing can be said of Africa and South America, where demand for fish is likely to keep growing. Europeans and North Americans can buy themselves a clear conscience with MSC-approved products, but that will not solve the world's overfishing problem. Plus, so-called "responsible" fisheries often make it possible to cover up the sale of other species caught in much less sustainable conditions: What is the point of ecocertifying cod destined for British fish and chips if most of the catch is made up of illicitly harvested small coastal sharks, many of which are endangered? These fraudulent practices are extremely common and largely facilitated by a deep-seated (and carefully maintained) laxity in the labeling and traceability of seafood. Jennifer showed that a full third of seafood in the United States was not correctly identified.[16] Many a restaurant customer would be furious to learn that their expensive wild grouper was in fact a farm-raised tilapia in disguise.

Jennifer's thesis didn't exactly thrill her colleagues at the MSC in London, but she was just getting started. She continued working on the issue, this time at the head of an international team, publishing an article in *Nature* that launched her onto the world stage.[17] Her tone was even sharper than before. According to Jacquet and her colleagues, the MSC's criteria are quite simply too lax and the environmental benefits they promise too nebulous. Indeed, many declining fisheries are now ecocertified, and the MSC team seems to care more about legal problems than environmental ones. Even more scandalous, reduction fisheries can now get the MSC label, a major reason for its declining credibility. "We no longer have to prove that extracting such huge quantities of fish, which form the basis of marine food webs, harms populations of large predatory fish, birds, and marine mammals—but the definition of 'sustainable' ought to include an 'ethical' obligation," says Frédéric Le Manach, one of Daniel's former students and now scientific director of BLOOM, addressing the problem of reduction fishing. "Catching perfectly edible fish, often in Africa or South America, to produce low-quality farmed fish for Westerners just doesn't make any sense. It's scandalous that the MSC label allowed itself to fall into that trap, and it's one of the many mistakes that will lead to its downfall."

"Daniel doesn't think that the individual has the power to change things as a consumer," continues Jennifer, who agrees "philosophically," but keeps to a strictly vegan diet. As an example, she points to the well-known case of ozone: after Sherwood Rowland and Mario Molina of the University of California discovered that fluorinated gas emissions from refrigerators were slowly eating away at the atmosphere, the powers that be did not wait for each citizen to "be reasonable" and buy a refrigerator that didn't emit fluorinated gas. No—they outlawed the production of the gas in question. In the same way, Daniel insists, countries should make sure fishing quotas are respected, stop subsidizing activities that are harmful for the marine environment, and establish protected marine reserves that are strictly off-limits to fisheries to allow fish populations to recover.

During a visit to Philippe Cury, looking out over the tuna-fishing port of Sète, Daniel continued reflecting on the measures that would need to be undertaken to resolve the marine biodiversity crisis and the environmental crisis as a whole. Together, Philippe and Daniel wrote *Mange tes méduses!* (Eat your jellyfish!), 180 pages of brilliant scholarship. Inspired by Daniel's beloved Darwin, the book begins with a long history of life, the biosphere, and humanity: "The planet-wide triumph of *Homo sapiens* is without question the most unusual and surprising event in the history of evolution[18]...While nature operates on natural cycles, humans operate on a unidirectional timeline (often called 'progress'), which is characterized by a rush forward and permanent expansion...[19] We relegate animals to the role of machines destined to fulfill our own needs."[20] Philippe Cury, who wrote the first act of this evolutionary tragedy, evokes a whole bestiary to communicate the wonder he feels when looking at the natural world, particularly the marine megafauna (large fish, turtles, birds, whales) that an "obstinate nature" forces to travel thousands of miles each year to feed and reproduce.

The mustachioed pair* tell the sad story of how humans have overexploited and destroyed ecosystems: "Three major phases are considered: first, that of hunter-gatherers who destroyed the large

---

* In the 1990s, Daniel traded in the beard of his youth for a more distinguished mustache.

animals that were readily available; the second is that of the domina-
tion of nature and the exhaustion of the soil through agriculture; and
finally, the most recent period in which technological advancements
have made it possible for man to systematically destroy renewable
resources... Humans are in such a hurry that they are ill-suited to the
slow cycles of nature, and prevent nature from renewing itself." [21]

This introduces the subject of overfishing, for which Daniel lays out
one of his favorite metaphors:

> The ocean illustrates how our interactions with the spoils of nature
> resemble a Ponzi scheme, named for the con man who invented it
> in the 1920s. His *modus operandi* was to promise incredible profits
> financed by an influx of capital which would be gradually invested.
> This type of financial structure always results in a speculative bub-
> ble and the collapse of the system in place... We believed that
> the natural capital of the oceans could produce a hundred million
> metric tons annually, but in fact, that is nothing more than a
> speculation on the world of living things, that we ourselves
> invented... The whole thing was brought into being by nothing
> less than a fisheries-industrial complex—an alliance between big
> fleet owners, lobbyists, lawmakers, and fisheries economists. [22]

Cury and Pauly remind their readers that members of this cartel
"guarantee themselves political influence and access to government
subsidies that are completely disproportionate to any contributions
they could hope to make to the GDP of advanced economies—in the
United States and France, this sector represents a smaller part of the
economy than the hair salons, and, in England, less than the lawn
mower industry."

Following this now-familiar overview of history and current
events, Cury and Pauly imagine possibilities for the future based on
their research and a lifetime of experience, though they also draw on
the UN's Millennium Ecosystem Assessment.* Their epilogue aims for

---

* millenniumassessment.org.

optimism, but it brings tears to my eyes: "Perhaps one day we will be able to recognize the beauty of life as a whole, without sentimentality, to be at ease with it and give up the idea of achieving supremacy over nature. The best scenario for the future that we can construct is the one imagined by Darwin: one in which humans live in nature and recognize the unity of all life and adjust their relationship to other living things accordingly. A change in the way we extract resources from the earth is necessary."[23]

Cury and Pauly conclude on an ambiguous note: "Our destiny is determined neither by heavy pessimism nor insipid optimism. Between all sorts of inconvenient truths and reassuring lies, we may yet manage to find our way toward a more sustainable and desirable society." Their text, which calls for an overall change of pace in our daily lives and in history,[24] touches on the ideologies of degrowth and simple living without citing them directly. Such a call to action would sound more coherent coming from someone like Pierre Rabhi* than from my two favorite fisheries scientists, whose carbon footprints are literally sky-high and who have been living at a hundred miles per hour for the last several decades.

---

* Pierre Rabhi (b. 1938): French environmentalist, writer, and farmer known for his philosophy of "happy sobriety." (TN)

FIRST LOVES,
FINAL BATTLES

"**D**ANIEL HAS A few recurring obsessions," Philippe Cury tells me. And it's true: when he received the International Ecology Institute's prestigious ECI prize in 2007, the German organization asked him to write a book for their Excellence in Ecology series. Everyone expected a handsome volume on overfishing—but he surprised them by proposing a two-hundred-page book titled *Gasping Fish and Panting Squids: Oxygen, Temperature and the Growth of Water-Breathing Animals*. Written during a stay in Bremerhaven immediately following Daniel's time with Philippe Cury in Sète in the spring of 2009, the book is a return to the PhD research he had done in Germany thirty years before. Daniel came back to his "oxygen story" because he had never been able to accept the larger scientific community's failure to notice his idea, nor its outright rejection by his physiologist colleagues.

Let us reiterate the key events of this story: in 1979, young Daniel wrote a densely theoretical doctoral thesis in which he argued that the growth and maximum size of fish were limited by the surface area of their gills, through which they absorb oxygen from the water. According to Pauly's theory, in fish and other water-breathing animals, the surface area of the gills (or other respiratory surface) grows more slowly—simply because it is a surface—than the oxygen requirement of these animals, which is a function of their volume. Oxygen supply *should* therefore be a major limiting factor for fish size. "Among my closest friends and colleagues, my apparent obsession with the relationship between oxygen and fish growth has become a friendly joke," recounts Daniel, "but I never stopped looking for evidence... and what I found was a mountain of information that corroborates my ideas, and which I knew I would have to publish when the time was right." Now that he had the necessary fame and seniority, the moment had come for Daniel to produce an improved version of his pet theory. His wife Sandra knew the oxygen story like the back of her hand—she had been there when Daniel first thought it up during their first year together in Kiel, on the edge of the Baltic—and she stood behind her husband, encouraging him to "never give up."

During the summer of 2009, Daniel got to work on a much-improved second version of his beloved theory, though he also took advantage of his sabbatical in Bremerhaven to renew ties with some old friends. He enjoyed a few of Petz Arntz's home-cooked meals and got together with his old traveling companion Walter Kühhirt, whom he hadn't seen since the 1970s. The two friends took long walks on the beaches of the North Sea and visited the island of Helgoland, a breath of fresh air that Daniel needed after his frenetic writing sessions. His text vigorously elaborates on the implications of his theory, pointing out that it is both simple and capable of explaining a large variety of biological phenomena. In particular, he demonstrates that fish with larger gills grow faster than the others, that fish under stress (whose oxygen requirements are higher) grow more slowly, and that fish who live in colder, more highly oxygenated water are generally larger.

Daniel's text is detailed, powerful, and contains a good many practical examples. "I surrounded my favorite hypotheses with a 'defensive ring' of arguments, similar to the moat around a castle," Daniel writes.[1] The work is an impressive one, but the editors of the Excellence in Ecology series remained cautious. Otto Kinne, the founder of the International Ecology Institute, writes, "Although it is too early to tell whether the theory underlying this book will become widely accepted, I warmly congratulate Daniel Pauly for the effort." Daniel managed to convince some of his colleagues, however, including the influential physiologist Hans-Otto Pörtner, but once again, his work was initially snubbed by the majority of fish physiologists, even at his own university in Vancouver. His message did not go over any better internationally: according to my colleague David McKenzie of the National Centre for Scientific Research (CNRS) in France, "That theory is based on a misunderstanding of fish physiology." Daniel fumed some, but he also knew that with the help of a new heavyweight ally he would win this round, too.

"I GREW UP in the countryside around Hong Kong," William Cheung tells me. I stop him right there: "Isn't the countryside around Hong Kong a concrete jungle next to the sea?" William laughs nonchalantly. "No, some parts of the river basin are protected to guarantee the city's water supply, or at least they were when I was a kid. I was actually surrounded by a protected environment with a view of the China Sea, where I had a lot of experiences with the natural world that really made an impression on me." The young William chose to study biology, and his master's thesis was directed by the energetic Yvonne Sadovy de Mitcheson, a London native and new professor at the University of Hong Kong who had also worked in Puerto Rico and the Bahamas. "Yvonne is very enthusiastic," William recounts. "She was the one who exposed me to Daniel's work in the 1990s. Like every marine biology student, I wanted to go scuba diving and count fish. Yvonne suggested something a little more ambitious—looking at how Hong Kong's fisheries and marine ecosystem had changed during the second half of the twentieth century."

William spent two and a half years assembling all the information available in various libraries, interviewing 150 fishermen, divers, and functionaries, making an Ecopath-type model, and reconstructing fishing statistics as best he could. This allowed him to identify marked overfishing starting in the 1960s and to propose better management solutions.[2]

"At the time, we weren't able to use Ecopath in Hong Kong, and Yvonne sent me to Vancouver for three months in 1999. That experience changed my life. I was exposed to all kinds of new ideas. Daniel was just beginning the Sea Around Us project, and there was this great sense of inspiration. After that, I knew I wanted to go back to UBC for my PhD, but I spent two years working for the WWF first." When he did return to Vancouver, his thesis director was Tony Pitcher, who helped him create statistical models to estimate the vulnerability of different fish species to overexploitation and extinction.

But Daniel was never far from this student who, he says, "had a real head for numbers." When William's PhD funding ran out in the mid-2000s, Daniel offered him a contract with the Sea Around Us, asking him to refine their distribution models for different fish species. William found himself "on the shoulders of giants," as he puts it—he benefited from both FishBase and the experience of colleagues Reg Watson, Chris Close, and Vicky Lam, who had already spent several years thinking about the best way to predict the distribution and abundance of fish exposed to both fishing and the warming of the oceans. Indeed, if the effects of overfishing are already well known, we are only just beginning to understand how climate change will affect fish populations. Oceanographers have known for a long time that warm waters are less productive—their beautiful, translucent blue color only betrays the sheer emptiness of these biological deserts. Fish find less food there, and the water is lower in oxygen, which is, of course, essential for their growth and survival.

Slowly but surely, William would figure out how to simulate the impact of global warming on the oceans and fisheries worldwide. He became an expert in this area and built an international reputation, publishing increasingly prestigious scientific articles that have been cited by thousands of his colleagues. In particular, he led the Sea Around Us team that, along with colleagues at Princeton, estimated what the distribution ranges will be in 2050 for more than a thousand species of fish and marine invertebrates exploited by fisheries all over the world.[3] To accomplish this feat, the researchers used the climate change scenarios published by the IPCC in 2007. Their seminal study showed that because of human-induced climate change, numerous marine species will soon disappear from the tropical and subtropical regions, which will become too hot, while the polar regions could find themselves invaded by species from lower latitudes. According to Cheung and his coauthors, this scenario could lead to a dramatic replacement of 60 percent of species currently present in these regions and have serious consequences for marine ecosystems and food security in tropical countries. In fact, the same team, still under William's leadership, published a second study shortly afterward indicating that the fisheries that will benefit the most from global warming between now and the year 2055 are in Norway, Iceland, Greenland, Alaska (United States), and Russia, while the biggest losses will be seen in the Pacific, China, and from Indonesia to Ecuador. The irony of the situation is that the Global North could actually be "rewarded" for their massive greenhouse gas emissions by a huge influx of fish in their waters.[4]

But is the "tropicalization" of the oceans already happening? Like any good biologist, William already knew that most species of fish cannot survive outside of a specific thermal zone (with the exception of large sharks and tuna). He therefore decided to calculate the average temperature of the water where each exploited species lives. "I spent a whole summer compiling the data," says William, who stayed at the office while everyone else went to the beach. With help from Daniel and Reg Watson, he processed data from the Sea Around Us's impressive stores for a thousand fish species living in fifty or so large marine ecosystems around the globe. His analysis revealed that the average temperature in the ranges of exploited fish had indeed increased by 0.7 °C (1.26 °F) since the 1970s.[5] This difference may seem small, but in an underwater world that is extremely sensitive to changes in temperature, such an increase is significant and clearly signals that warming is already under way.

William completed and improved his statistical models of the distribution of marine species by taking fish physiology into account. "In the summer of 2009, I was in the UK, and I took a side trip to see Daniel in Bremerhaven," William recalls. "He was neck-deep in his 'oxygen story,' and he had me read his doctoral thesis, which really impressed me." William integrated an additional constraint into his statistical models, taking into account the oxygen content of the water and its effect on fish growth and size. Daniel was overjoyed: "I gave William the keys to a Ford Fiesta and he came back with a Rolls Royce!" Indeed, William had illustrated Pauly's theory beautifully, calculating that for six hundred fish species, the maximum size of individuals could fall by 14 to 24 percent by 2050 as a result of global warming.[6]

But the scientific controversy remained intense. In March of 2017, Sjannie Lefevre and Göran Nilsson of the University of Oslo, along with David McKenzie of the CNRS, published an opinion piece in the journal *Global Change Biology*, titled "Models Projecting the Fate of Fish Populations Under Climate Change Need to Be Based on Valid Physiological Mechanisms."[7] According to these authors, gill lamellae, which transfer oxygen into the bodies of fish, are like the letters on the pages of a book, whose numbers can increase with the book's volume. Thus, gills' surface area growth and their capacity to extract oxygen from water go hand in hand with the fish's volume and its metabolic requirements: the availability of oxygen should, therefore, not limit the growth or maximum size of animals that breathe underwater.

Daniel's response, also signed by William, came quickly, and reading it, you can't help but feel sorry for the young Norwegians who defied the old master.[8] In energetic prose, Daniel advances a slew of arguments in favor of his theory. "I've always had trouble getting people to understand how gills work," he tells me, "but I've finally found the right metaphor: they're like the radiator in a car, the surface of which limits the capacity of the motor"—unlike letters in a book. More scientifically, Daniel and William explain that gills are organized into lamellae, and that a given gill surface can only hold so many lamellae.

As he regales me with the details of his most recent game of academic ping-pong in the fall of 2017, I can sense how much Daniel enjoys these public skirmishes. But the party wasn't over yet: the other team parried soon after.[9]

Other battles rage on as well. The most striking pits Daniel against his American colleague Ray Hilborn. Yet another mustachioed scientist, he is only eighteen months younger than Daniel. Like him, he became a fisheries scientist by chance in the 1970s, and like him, he is a big name in the field, known for his many publications, prestigious works, and the awards he has won. Hilborn grew up in California and studied under Carl Walters at UBC. Since 1987, he has been a professor at the University of Washington in Seattle, only about 120 miles south of Vancouver. Well-spoken, Ray describes his first encounter with Daniel in 1988 as follows: "The first time we met, flying from Halifax to Toronto after a conference, he spent the flight outlining faults in my presentation and listing in great detail all the work going on around the world that I had not referred to, and in truth, was unaware of. I learned that when in need of a good critic, sit next to Daniel Pauly." This anecdote can be found in the preface Hilborn wrote for Daniel's 1994 book *On the Sex of Fish and the Gender of Scientists: A Collection of Essays in Fisheries Science*. Hilborn's text is teasing but mostly laudatory, even describing Daniel as "the most widely cited fisheries scientist of his generation" (and therefore more cited than Ray himself, which is still the case today). At that time, Hilborn and Pauly were mostly in agreement about what was wrong with fisheries and about the need to improve their management. They strongly disagreed, however, about how to handle the statistics. Daniel and his team developed ELEFAN and other methods suited to the sometimes difficult research conditions in the Global South, but such approaches were often criticized by colleagues from the North, including Hilborn. Ray is, in fact, the author of the famous quote, "The Prophet Daniel is among those excluded . . . Daniel must toil in infernal heat, deprived of holy catch-at-age data and armed only with a thermometer"[10]—though he probably wrote it without knowing that the target of his ire had grown up an outcast.

Ray Hilborn, a white North American conservative, and Daniel Pauly, a mixed-race alter-globalist with Marxist leanings, are worlds apart. But they share the same pride and ambition, as well as a quarrelsome nature. They crossed swords more frequently as time went by, and the tension mounted even as Daniel's notoriety increased—in the early 2000s, the press crowned him "the Fisher King."[11] But Daniel was not Hilborn's only target: "Ray Hilborn and Carl Walters had been criticizing Ransom Myers for a long time," Boris Worm tells me. The Hilborn-Walters duo were also some of Boris's most virulent detractors in 2006 when he predicted the collapse of marine fisheries by 2048.[12] But Boris chose to respond diplomatically: he swallowed his pride and invited Ray to a synthesis workshop on global fisheries. The results, which aimed for a reassuring tone, were published in *Science* three years later.[13] Boris invited Daniel as well, but he declined in order to "avoid a pissing contest" with Hilborn. Instead, he sent Reg Watson and Dirk Zeller to represent their team.

"Since we had worked together, it was harder for Ray to come after me," Boris told me with a laugh, "so he threw everything he had at Daniel." When I interview him at his farm in Washington State, Ray describes himself as a rationalist who is generally skeptical about his colleagues' work and who promotes a "pragmatic" approach to managing marine ecosystems. He freely admits that the methods and databases (FishBase, Ecopath) developed by Daniel and his team are very useful to the research community, but he knocks their reconstructions of fishing statistics as "an absolute waste of time" and marine reserves as the con of the century: "Well-managed fisheries do not need marine protected areas," Ray declares. He is, therefore, not the least bit upset about Donald Trump's plans to open up the largest marine protected area in the world, located in the middle of the Pacific, to American fisheries. Ray insists that the oceans need to be exploited on an industrial scale in order to feed humanity, arguing that such exploitation would be much less harmful to the planet than intensive agriculture. To achieve that end, he finds it perfectly reasonable to "massively overfish" large, predatory fish species in order to increase the yield of smaller fish. Hilborn considers the Paulyesque vision of the oceans to be "catastrophist" and says that "a major challenge for fisheries management is posed by environmentalism."[14] According to Ray, Daniel and the NGOs who rely on his work are using the fear created by their apocalyptic vision of the state of the ocean to drum up more funding.

Funding also turns out to be an interesting topic in Ray's case: "My attitude is, actually, I'd like to take money from everybody," Ray tells me. "I've received funding from environmental NGOS... I've been very successful in these last few years at getting increasing amounts of money from the fishing industry." This last source seems to make up the majority, however. In the spring of 2016, at the end of a European conference tour, Ray made headlines in a less-than-flattering way. The French newspaper *Le Monde* published a full page on him titled "Ray Hilborn, a fishing expert in the eye of the storm."[15] The article declared that "Greenpeace has published compromising documents on an influential American fisheries scientist's conflicts of interest." In particular, they accused Hilborn of "taking 3.56 million dollars in funding from fishing-related interests, or 22 percent of his total funding for research and knowledge sharing" for the period between 2003 and 2015.* "According to Greenpeace," the article continues, "no less than sixty-nine organizations with ties to the fishing sector—businesses, foundations, professional associations—have supported the famous scientist's work... In addition to this funding, the scientist has also received personal payments."

What Greenpeace denounced was not the fact that Hilborn had used funds from those sources, but that he had not systematically declared a conflict of interest in his scientific publications, in accordance with standard practice. As an example, *Le Monde* cites an episode in 2010 where "Ray Hilborn wrote a letter to *Nature* in support of the Marine Stewardship Council (MSC), an organization whose goal is to guarantee that fisheries are exploited sustainably. The well-known label had been called into question by certain researchers. Two years later, he authored, alongside several other scientists, a study that was favorable to the MSC in the journal *PLoS One*. In both cases, he failed to declare that, between 2008 and 2010, he had worked as a paid consultant for one of the certification companies used by the MSC."

---

* Ray Hilborn's research had therefore received a total of 1.2 million dollars in funding each year.

The association France filière pêche,* which had invited Ray Hilborn to Europe, was embarrassed. According to *Le Monde*, "the association states that they only compensated the researcher for his travel expenses and those of his spouse."[16] Eighteen months later, I ask Ray to tell me about the events that followed. He says that the journal editors had looked into it, but concluded that he had done nothing wrong, and, he adds, "The University of Washington said I hadn't violated any rules."

Daniel and Ray are both septuagenarians now, but their intense rivalry confers a sort of eternal youth, and just a hint of mischievousness, on both of them. Ray turns out to be a good sport after all, ending our interview with the following anecdote: "Several years ago I gave the Peter Larkin Lecture at UBC, and Daniel introduced me by saying that he thought my coming to UBC was like the Pope going to Mexico and telling people they didn't need birth control—this in reference to my skepticism of the benefits of marine protected areas in many places."

---

* An interprofessional association that includes members from all sectors of the French maritime fishing industry.

EPILOGUE

I T HAS TAKEN me two years to write the biography of Daniel Pauly, which I completed while also working on my own research. In that time, Daniel has added fifty titles to a list of publications that already included five hundred, leading me on a wild goose chase in which new texts appeared faster than I could read them. His work, already cited 64,000 times, has since attracted 13,000 new citations, confirming his status as one of the most influential authors in the history of the marine sciences. He also received seven more distinctions between 2016 and 2017 (making forty in all), including the French Légion d'honneur. Radio-Canada named him scientist of the year in 2016, and he acquired Canadian nationality. His new passport only appeared, however, after a few administrative hurdles: Daniel had to provide precise dates of all the times he had exited or entered Canada over the previous ten years, along with documents to back them up.

His seventy-second birthday come and gone, Daniel doesn't seem keen on retirement, even if, as Sandra points out, "When he insists on working over the weekend, he's exhausted by Wednesday." Daniel doesn't direct a dozen PhD students at a time anymore, as he did in the 1990s and 2000s, though he does teach and give conferences all over the world. In the last two years, he has been to India, China, Australia, New Zealand, New Caledonia and Tahiti, Norway, England, Germany, the Netherlands, Belgium, Switzerland, France, the United States, Mexico, Brazil, the Galápagos, and South Africa. "With his profile, Daniel should be named a goodwill ambassador for the United Nations," his friend Kostas Stergiou tells me.

But superheroes are solitary, and Daniel is no exception. During the months I spent at his institute in Vancouver, I was surprised to find him isolated. After many years of sheltering in prefabricated buildings, the Institute for Oceans and Fisheries has moved into a shiny new building, a concrete structure with a large entryway and suspended walkways worthy of a shopping mall. The students are crowded into a forest of cubicles, while Daniel and the other researchers occupy individual hundred-square-foot offices. I'm just a friendly bird lover, and everyone is very cordial with me, but the tension is palpable. The leading lights of marine science don't speak to each other, are hardly ever around for coffee breaks, and their students don't mix. Daniel explains to me that a "process of alienation" led him to distance himself from most of his colleagues and some members of his own team. I would add that he has long overshadowed other scientists of his generation, and that his almost-Stakhovite work ethic has generated some resentment. "Daniel has always put his work before everything," Philippe Cury tells me, "including friendship." Jennifer Jacquet adds, "Daniel doesn't need other people in order to function." And it's true that he mostly lives his life in the hundred square inches of his computer screen and on the pages of the books he continues to devour at lightning speed.

But everyone agrees that since his stroke, Daniel has become more tolerant, affectionate, and empathetic. And this still-unwrinkled septuagenarian is lucky—Sandra watches over him daily, Deng Palomares is as loyal as ever, and Cornelia Nauen keeps an eye on him from afar. When he is at the office in Vancouver, Sandra joins him for lunch every few weeks with William Cheung, Rashid Sumaila, and their wives. "We go for dim sum," says Daniel, for whom the pleasures of the table are still of capital importance. Over lunch, he regales his friends with a few monologues on the most recent scientific discoveries he has spotted in *Science* and *Nature*, which he still consults religiously. These rare social gatherings give Daniel much-needed breathing room, a port in a sea of troubles.

In 2014, the Pew Charitable Trusts cut off the funding that had made the Sea Around Us a model for fisheries research around the world. The Paul G. Allen Family Foundation* picked up the slack in 2015, but the Sea Around Us found itself with a much lower budget and a lot less control over its activities. Deng Palomares made frequent trips to the foundation's headquarters (fortunately located in nearby Seattle) to defend her team's interests and explain that scientific research doesn't always advance at the speed of a microprocessor. Despite her efforts, though, the Sea Around Us's financial situation worsened in 2016, and Daniel was forced to let many of his teammates go. "There were fifteen of us," Deng says nervously, "but now there will just be me, Daniel, and three assistants. Daniel hasn't played the political game in Canada," she adds, "which says something about his integrity, but it also comes at a cost for people like me, who still don't have permanent positions."

Daniel seems to be running out of reasons to stay in Vancouver, that artificial, bourgeois paradise, where his team is beginning to shrink. His children live far away, as does his French family, so he often hits the road, thanks to an impressive network of allies all over the world who are always happy to take him in. "I was talking to Daniel on the phone, and he told me he didn't want to go to the office anymore," Jennifer Jacquet recalls. "I could not believe it—I'd never heard Daniel say he doesn't want to go to work. I spoke to my boss, and I said, 'Do you think we could bring Daniel here?'" No sooner said than done— Daniel and Sandra spent three months in Manhattan in the fall of 2016.

I join them there for Thanksgiving, rushing out of Kennedy Airport, determined not to be late for the dinner Sandra has planned—an Arkansas-style chateaubriand. We are happy to see each other, but Sandra warned me ahead of time: "Daniel is very tired—recent events have taken a lot out of him." Donald Trump had been elected president of the United States two weeks earlier. "It's like a kick in the pants," Daniel tells me. Always a loyal reader of the *New York Times*, he has stopped paying attention to the news. "There aren't any bright futures anymore," the Baby Boomer concludes sadly. Daniel has fallen

---

* Established by the cofounder of Microsoft.

back on different reading materials, and I notice during my visit that he has made a valiant effort to fill up the bookcase in their temporary home, mostly with science fiction. "If it's set in space with big space-ships, I love it!" he tells me, his eyes sparkling momentarily.

I crash on the Paulys' couch, setting my alarm for 6:00 AM—which just leaves me time to shake myself awake before Daniel strolls into the living room, laptop under his arm. He puts the kettle on ("My modest contribution to breakfast," he grumbles) and dives into his email. As he wolfs down his bacon, toast, and fruit, Daniel launches into a con-versation on string theory before moving on to astrophysics. If you're planning to eat breakfast with Daniel Pauly, I suggest doing some intel-lectual warm-up exercises first. Our conversation continues as the day wears on, with an outing to Daniel's favorite Italian restaurant, where he shows me how to hold a piece of pizza in one hand "like a real New Yorker," then a walk among the famous chess players of Washington Square Park.

"In three months, Daniel and Sandra barely set foot outside of that one square kilometer," sighs Jennifer Jacquet. Daniel wrote yet another book, helped out with a study showing the link between illegal fishing and modern slavery,[*] and taught classes for American students who were only too happy to meet a living legend. Then the lifelong wan-derers set off for California to spend the holidays with Sandra's family, where they were joined by Angela, Ilya, and their partners. "And it's crazy, he still does not fly business," Jennifer exclaims, "despite his height, his age, and his health." I tell her that I will quote her on that, but I don't think anyone will believe me.

In 2017, the Paulys were invited to Australia for a happy occasion: Dirk Zeller, Daniel's teammate for the last twenty years and the artisan

---

[*] According to the study in question, residents of the European Union are thirteen times more likely to eat fish caught by enslaved fishers when the products are imported from developing countries. This kind of modern slavery is common on African, Asian, South American, and even Russian fishing boats. See: David Tickler, Jessica J. Meeuwig, Kath-arine Bryant, Fiona David, John A. H. Forrest, Elise Gordon, Jacqueline Joudo Larsen, Beverly Oh, Daniel Pauly, Ussif R. Sumaila, and Dirk Zeller, "Modern Slavery and the Race to Fish," *Nature Communications* 9, no. 1 (2018): 4643.

behind many a statistical reconstruction, finally landed a professorship at the University of Western Australia. "It was about time—Dirk is well past fifty," comments Daniel, who enjoyed the indolent, Mediterranean atmosphere of Perth. "They loaned me a very nice apartment, and Sandra was able to come along." Daniel's morale slowly improved, especially when he learned that Dirk Zeller would be opening a local branch of the Sea Around Us thanks to the University of Western Australia, which had agreed to become one of the funders of FishBase. Better yet, he met a millionaire philanthropist who knows a few billionaires determined to do something good for the oceans. "The Sea Around Us's financial situation will be better in 2018," he tells me, radiant. In Vancouver, Deng can breathe a little easier—they might be able to bring back some of her colleagues who were let go in 2016.

Things are looking up for the fish as well, at least in some regions. I called Didier Gascuel to check in on the health of French fisheries. Didier explains that the patient had been sickest in the late 1990s: overfishing was everywhere, yields were down despite the subsidies, and the increased price of fuel made the sector even less profitable. Quite a few fishers threw in the towel, a large number of big fishing boats were retired, and the pressure eased. As a result, French fisheries have recovered somewhat in the last few years, particularly in the English Channel and on the Atlantic coast, even if the situation in the Mediterranean, where the sardine stocks have collapsed, remains worrisome. The European Union has not been as effective when it comes to managing its fisheries as certain Northern European countries. Norway is the best in its class for marine fisheries, though it does have one small advantage: climate change is causing the North Atlantic to warm noticeably, and fish in the Norwegian and Barents Seas love the milder temperatures. Overall, if we are to believe Didier Gascuel, countries in the Global North are doing their best to rebuild their fisheries and improve their management. Ray Hilborn and his gang are not, therefore, entirely wrong when they declare the state of the world's fisheries satisfactory—they are just blinded by a situation unique to the United States. Further south, the disaster continues to unfold off the coast of

Africa and, generally speaking, in all the tropical countries, especially in Southeast Asia.

That is why Daniel continues to insist, as he always has, that it is essential to look at the future of fish and marine ecosystems on a global scale. The international NGO Oceana is the number-one advocate of this point of view, backed by powerful American organizations including the Pew Charitable Trusts and the Rockefeller Foundation. Daniel's ties to this sphere are close and long-standing—he is, quite logically, on Oceana's board. Andy Sharpless, Oceana's executive director, explains their approach: "We went to see the Chinese ambassador in Geneva . . . We said, 'We have a bad history, as a planet, in managing the oceans; they are collapsing.' And he said, 'We have a billion people in China, they're very hungry, we're gonna feed them.'"* Andy took the blow in stride and came back swinging in 2013 with a book titled *The Perfect Protein: The Fish Lover's Guide to Saving the Oceans and Feeding the World.*† His arguments are largely based on the work of Daniel Pauly, William Cheung, and Rashid Sumaila: By 2050, there will be nine billion people on Earth. Agriculture will still provide much of our food, though often at the cost of biodiversity and terrestrial ecosystems. The environmental impact of well-managed marine fisheries, on the other hand, is negligible and the nutritional quality of fish far superior to that of other kinds of meat. Currently, 750 million poor people live in countries with access to marine resources, while our overfished oceans can barely feed 450 million people per year. If we managed the world's marine resources more effectively, we could feed 700 million people, and if we eliminate reduction fisheries,‡ that figure could rise to over a billion people. The actions that would need to be taken to rebuild fish stocks are well known: quotas must be respected, reproduction sites protected, and marine reserves created; and fisheries must

---

* Oceana, "Andrew Sharpless," oceana.org/about-oceana/people-partners/executive-committee/andrew-sharpless.
† Coauthored with Suzannah Evans, with a preface by Bill Clinton.
‡ As we have seen already, reduction fisheries take small pelagic fish that could be eaten by humans and use them to feed farmed fish.

be made more selective in order to reduce collateral damage. Andy Sharpless has said it over and over again: "The basic facts of what you need to do to produce an abundant ocean have been proven . . . you get the fish back, [and] you get them back in five or ten years—this is not some academic abstraction."

Environmental NGOs have lobbied successfully in countries all over the world. I asked Didier Gascuel about the Mediterranean bluefin tuna, the subject of an impressive battle between NGOs, politicians, and fishing executives that has led to stricter regulations. According to Didier, "The stock is coming back quickly, that's almost certain. The NGOs saved fishing by forcing the industry to take the necessary steps. But the fishing sector is as insatiable as ever, and the most recent quota increases are ill-advised."

The myth of inexhaustible ocean resources has finally been shattered: overfishing is a media phenomenon, so obvious it cannot possibly be ignored. Daniel has become the ocean's great whistleblower, an exceptional individual who—in a scientific community shackled by economic interests—dared to stand up and tell the public about the impending doom of fish populations, even at the risk of angering many of his colleagues. But life has not made Daniel Pauly "like everybody else." Exclusion gave him all the courage he needed.

APPENDICES

IMPORTANT DATES
IN THE LIFE OF
DANIEL PAULY

MAY 2, 1946   Daniel is born in Paris, the son of Renée Clément and Winston McLemore

SUMMER 1948   Daniel is entrusted to a Swiss family from La Chaux-de-Fonds

JULY 1963   Expelled from high school, Daniel leaves La Chaux-de-Fonds for Germany

AUTUMN 1964   Daniel finds a job in a Wuppertal factory and returns to his studies

JANUARY 1967   Called up for military service, Daniel returns to France and reunites with Renée

SPRING 1969   *Abitur* (German baccalaureate) in Wuppertal

SUMMER 1969   First visit to the United States, where Daniel meets Winston McLemore

SEPTEMBER 1969   Daniel starts college in Kiel

1971   First research expedition in Africa (Ghana)

1974   Master's degree at IfM in Kiel

JUNE 1975	Daniel leaves for Indonesia
MAY 1977	Ilya Pauly is born
DECEMBER 1978	Marriage to Sandra Wade
MAY 1979	Doctorate at IfM in Kiel
JULY 1979	Daniel takes a job at ICLARM; the Pauly family moves to Manila
MARCH 1981	Angela Pauly is born
1994	Daniel becomes a professor at the University of British Columbia in Vancouver
1995	First publication in *Nature*, *"Primary Production Required to Sustain Global Fisheries"*
1998	First publication in *Science*, "Fishing Down Marine Food Webs"
1999	Launch of the Sea Around Us project
2003	Daniel is named one of the 50 most influential scientists in the world by *Scientific American*
FEBRUARY 2005	Daniel has a stroke
JUNE 2005	International Cosmos Prize (Japan)
2011	Grand prix of the Société française d'écologie (French Ecological Society)
2016	Prince Albert I medal (Monaco)
2017	Scientist of the Year 2016 (Canada)
2017	Chevalier of the Légion d'honneur (France)

LIST OF ABBREVIATIONS
AND ORGANIZATIONS

BASIC: Beginner's All-purpose Symbolic Instruction Code, a computer programming language

BLOOM: A nonprofit foundation for marine conservation

BP: Before Present

CGIAR: Consultative Group for International Agricultural Research

CNRS: Centre national de la recherche scientifique (French National Centre for Scientific Research)

CRODT: Centre de recherches océanographiques de Dakar-Thiaroye (Oceanographic Research Center of Dakar-Thiaroye)

CSIRO: Commonwealth Scientific and Industrial Research Organisation

DDE: Direction départementale de l'équipement (the authority formerly responsible for transportation and infrastructure in France, it was replaced in 2006 by the Directions départementales des territoires)

DDT: Dichlorodiphenyltrichloroethane

DKP: Deutsche Kommunistische Partei (German Communist Party)

ECOPATH: A free suite of computer programs for modeling ecosystems, initially created by Jeffrey Polovina of NOAA, but later developed by the Fisheries Centre at the University of British Columbia in Vancouver (now the Institute for Oceans and Fisheries)

ECOSIM: Individual-based Ecosystem Simulation Module

ELEFAN: Electronic Length Frequency Analysis

ENSO: El Niño Southern Oscillation

EWE: Ecopath with Ecosim

FAO: United Nations Food and Agriculture Organization

FISHBASE: Digital encyclopedia of fish species

FSC: Forest Stewardship Council (environmental certification designed to promote the sustainable management of forests worldwide)

GEOMAR: Helmholtz-Zentrum für Ozeanforschung Kiel (new name of the IfM)

GTZ: Deutsche Gesellschaft für technische Zusammenarbeit (German Agency for Technical Development and Cooperation)

ICES: International Council for the Exploration of the Sea

ICLARM: International Center for Living Aquatic Resources Management

IfM: Institut für Meereskunde (Oceanographic Institute) in Kiel, Germany (in 2004, the IfM was renamed GEOMAR)

IFREMER: Institut français de recherche pour l'exploitation de la mer (French Research Institute for Exploitation of the Sea)

IMARPE: Instituto del Mar del Perú (Marine Institute of Peru)

INRA: Institut national de la recherche agronomique (National Institute for Agricultural Research in France)

IPCC: Intergovernmental Panel on Climate Change

IRD: Institut de recherche pour le développement (Research Institute for Development in France); in 1998, the IRD replaced ORSTOM

MAVA: A nature foundation founded by French ornithologist Luc Hoffmann that is very active in Africa

MSC: Marine Stewardship Council (environmental certification designed to promote the sustainable management of marine resources)

NAACP: National Association for the Advancement of Colored People (United States)

NASA: National Aeronautics and Space Administration (United States)

NOAA: National Oceanic and Atmospheric Administration (United States)

NPA: New People's Army (a branch of the Philippine communist party)

OAK FOUNDATION: A consortium of philanthropic organizations headquartered in Geneva

ORSTOM: Office de la recherche scientifique et technique outre-mer (Overseas Scientific Research and Development Office), a French organization replaced in 1998 by the IRD

PEW CHARITABLE TRUSTS: American nonprofit organization that works to improve public policy, keep the public informed, and invigorate civic life

PNAS: *Proceedings of the National Academy of the Sciences* (United States)

RAF: Rote Armee Fraktion (Red Army Faction); a German communist organization also known as the Baader-Meinhof Gang

RAMSAR CONVENTION: Established by UNESCO in 1971, the convention's goal is to protect wetlands around the world

SEALIFEBASE: Digital encyclopedia of marine life

SEA SHEPHERD: Sea Shepherd Conservation Society; a non-governmental organization that works to protect marine ecosystems

SIAP: Système d'information et d'analyse des pêches (System for Information and Analysis of Fisheries)

UBC: University of British Columbia

UNESCO: United Nations Educational, Scientific and Cultural Organization

WWF: World Wide Fund for Nature

NOTES

FROM OCEANOGRAPHY
TO FISHERIES BIOLOGY

1. Mark Kurlansky, *Cod: A Biography of the Fish That Changed the World* (New York: Penguin Books, 1998).

DANIEL'S FIRST
AFRICAN EXPERIENCE

1. Jean de la Fontaine, "The Animals Seized With the Plague," in *La Fontaine's Fables,* trans. Robert Thomson (Paris: Chenu, 1806).
2. Daniel Pauly, "Report on the U.S. Catfish Industry: Development, Research, Production Units, Marketing and Associated Industries," in *Neue Erkenntnisse auf dem Gebiet der Aquakultur,* vol. 16, *Arbeiten des Deutschen Fischereiverbandes,* ed. K. Tiews (Hamburg: Deutscher Fischerei-Verband, 1974), 154–67. [In German.]

DEVELOPMENT AID
IN INDONESIA

1. Ludwig von Bertalanffy, *Theoretische Biologie—Zweiter Band: Stoffwechsel, Wachstum* (Bern: A. Francke Verlag, 1951).

2. Daniel Pauly, "On the Interrelationships Between Natural Mortality, Growth Parameters, and Mean Environmental Temperature in 175 Fish Stocks," ICES *Journal of Marine Science* 39, no. 2 (1980): 175–92.
3. Ray Hilborn, "Current and Future Trends in Fisheries Stock Assessment and Management," *South African Journal of Marine Science* 12, no. 1 (1992): 975–88.

BIRTH OF A CAREER
IN THE PHILIPPINES

1. Robert E. Ricklefs, *Ecology* (New York: Chiron Press, 1973).
2. Daniel Pauly, "Theory and Management of Tropical Multispecies Stocks: A Review, With Emphasis on the Southeast Asian Demersal Fisheries," ICLARM Studies and Reviews 1 (1979).

SAN MIGUEL BAY AND
THE SOCIAL DIMENSION
OF FISHERIES

1. Ian R. Smith and Daniel Pauly, "Resolving Multigear Competition

in Nearshore Fisheries," ICLARM *Newsletter* 6, no. 4 (1983): 11–18.

2. Daniel Pauly, "Some Definitions of Overfishing Relevant to Coastal Zone Management in Southeast Asia," *Tropical Coastal Area Management* 3, no. 1 (1988): 14–15.

3. Thomas Malthus, *An Essay on the Principle of Population* (London: J. Johnson, 1798).

4. Daniel Pauly, "On Malthusian Overfishing," *Naga: The ICLARM Quarterly* 13, no. 1 (1990): 3–4.

5. Pauly, "Some Definitions of Overfishing."

6. Daniel Pauly, "Major Trends in Small-Scale Marine Fisheries, With Emphasis on Developing Countries, and Some Implications for the Social Sciences," *MAST* 4, no. 2 (2006): 7–22.

7. Daniel Pauly, "Rebuilding Fisheries Will Add to Asia's Problems," *Nature* 433, no. 457 (2005), doi.org/10.1038/433457a.

A PACIFIC HEROINE

1. Siebren C. Venema, Jörgen M. Christensen, and Daniel Pauly, "Training in Tropical Fish Stock Assessment: A Narrative of Experience," in *Contributions to Tropical Fisheries Biology*, ed. Siebren C. Venema, Jörgen M. Christensen, and Daniel Pauly, FAO Fisheries Report 389 (1988): 1–15.

2. Jorge Csirke, John F. Caddy, and S. Garcia, "Methods of Size-Frequency Analysis and Their Incorporation in Programs for Fish Stock Assessment in Developing Countries," in *Length-Based Methods in Fisheries Research*, ed. Daniel Pauly and G. R. Morgan, ICLARM Conference Proceedings 13 (1987).

FISH STORIES IN PERU

1. Daniel Pauly and Isabel Tsukayama, "On the Seasonal Growth, Monthly Recruitment and Monthly Biomass of the Peruvian Anchoveta From 1961 to 1979," in *Proceedings of the Expert Consultation to Examine Changes in Abundance and Species Composition of Neritic Fish Resources*. San José, Costa Rica, 18–29 April 1983, ed. G. D. Sharp and J. Csirke, FAO Fisheries Report 291, vol. 3 (1983): 987–1004.

2. Reuben Lasker, "The Relations Between Oceanographic Conditions and Larval Anchovy Food in the California Current: Identification of Factors Contributing to Recruitment Failure," *Rapports et Procès-Verbaux des Réunions Cons. Int. Explor. Mer.* 173 (1978): 212–30.

3. Daniel Pauly and Isabel Tsukayama, eds., *The Peruvian Anchoveta and Its Upwelling Ecosystem: Three Decades of Change* (Callao, Peru: Instituto del Mar del Perú, 1987), 335.

4 Daniel Pauly, "Managing the Peruvian Upwelling Ecosystem: A Synthesis," in Pauly and

Tsukayama, *The Peruvian Anchoveta and Its Upwelling Ecosystem*, 325.

5. Daniel Pauly, Peter Muck, Jaime Mendo, and Isabel Tsukayama, eds., *The Peruvian Upwelling Ecosystem: Dynamics and Interactions*, ICLARM Conference Proceedings 18 (1989).

6. Arnaud Bertrand, Renato Guevara-Carrasco, Pierre Soler, Jorge Csirke, and Francisco Chavez, eds., "The Northern Humboldt Current System: Ocean Dynamics, Ecosystem Processes, and Fisheries," *Progress in Oceanography* 79, no. 2–4 (October–December 2008): 95–412.

7. Pierre Fréon et al., "Interdecadal Variability of Anchoveta Abundance and Overcapacity of the Fishery in Peru," *Progress in Oceanography* 79, no. 2–4 (October–December 2008): 401–12.

8. Alan R. Longhurst, *Ecological Geography of the Sea* (Cambridge, MA: Academic Press, 1998).

9. Alan R. Longhurst and Daniel Pauly, *Ecology of Tropical Oceans* (Cambridge, MA: Academic Press, 1987).

10. Daniel Pauly, "Why Squid, Though Not Fish, May Be Better Understood by Pretending They Are," *South African Journal of Marine Science* 20, no. 1 (1998): 47–58.

NATURE IN A BOX

1. Knud P. Andersen and Erik Ursin, "A Multispecies Extension to the Beverton and Holt Theory of Fishing, With Accounts of Phosphorus Circulation and Primary Production," *Medd. Dan. Fisk. Havunders* 7 (1977): 319–435.

2. Taivo Laevatsu and Herbert A. Larkins, eds., *Marine Fisheries Ecosystem: Its Quantitative Evaluation and Management* (England: Fishing News Books, 1981).

3. Jeffrey J. Polovina, "Foreword: The First ECOPATH," in *Trophic Models of Aquatic Ecosystems*, ed. V. Christensen and D. Pauly, ICLARM Conference Proceedings 26, 1993.

4. Polovina, "Foreword: The First ECOPATH."

5. Daniel Pauly, "On the Sex of Fish and the Gender of Scientists," *Naga: The ICLARM Quarterly* 12, no. 2 (1989): 8–9.

6. William E. Odum and Eric J. Heald, "The Detritus-Based Food Web of an Estuarine Mangrove Community," *Res. Chem. Biol. Estuar. Syst.* 1 (1975): 265–86.

7. Sylvain Bonhommeau, Laurent Dubroca, Olivier Le Pape, Julien Barde, David M. Kaplan, Emmanuel Chassot, and Anne-Elise Nieblas, "Eating Up the World's Food Web and the Human Trophic Level," *Proceedings of the National Academy of Sciences* 110, no. 51 (2013): 20617–20.

8. Robert E. Ulanowicz, *Growth and Development: Ecosystems Phenomenology* (New York: Springer, 1986).

9. Villy Christensen and Daniel Pauly, "ECOPATH II–A Software for Balancing Steady-State Ecosystem

Models and Calculating Network Characteristics," *Ecological Modelling* 61, no. 3–4 (1992): 169–85.

10. Robert E. Ulanowicz, "Foreword: Inventing the Ecoscope," in *Trophic Models of Aquatic Ecosystems*, ed. V. Christensen and D. Pauly, ICLARM Conference Proceedings 26, 1993.

11. Christensen and Pauly, *Trophic Models of Aquatic Ecosystems*.

12. Daniel Pauly, *Méthodes pour l'évaluation des ressources halieutiques*, trans. J. Moreau (Toulouse, France: Collection Polytech de l'INP de Toulouse, 1997).

13. Maria Lourdes Distor Palomares, "La consommation de nourriture chez les poissons: Étude comparative, mise au point d'un modèle prédictif et application à l'étude des réseaux trophiques," (doctoral thesis, Toulouse INP, 1991).

14. Carl J. Walters, *Adaptive Management of Renewable Resources* (New York: MacMillan, 1986).

15. Ray Hilborn and Carl J. Walters, *Quantitative Fisheries Stock Assessment and Management* (New York: Chapman and Hall, 1991).

16. Carl Walters, Villy Christensen, and Daniel Pauly, "Structuring Dynamic Models of Exploited Ecosystems From Trophic Mass-Balance Assessments," *Reviews in Fish Biology and Fisheries* 7, no. 2 (1997): 139–72.

17. Chiara Piroddi, Marta Coll, Camino Liquete, Diego Macias, Krista Greer, Joe Buszowski, Jeroen Steenbeek, Roberto Danovaro, and Villy Christensen, "Historical Changes of the Mediterranean Sea Ecosystem: Modelling the Role and Impact of Primary Productivity and Fisheries Changes Over Time," *Scientific Reports* 7 (2017).

FOR ALL THE FISH IN THE WORLD

1. Daniel Pauly and Rainer Froese, "FishBase: Assembling Information on Fish," *Naga: The ICLARM Quarterly* 9, no. 4 (1991): 10–11.

2. Robert A. McCall and Robert M. May, "More Than a Seafood Platter," *Nature* 376 (1995): 735.

3. Nicolas Bailly, "Why There May Be Discrepancies in the Assessment of Scientific Names Between the Catalog of Fishes and FishBase Version 2," WorldFish Center—FishBase Consortium—FishBase Information and Research Group, Inc., May 6, 2010, fishbase.us/Nomenclature/FBCofFNames.php.

4. Maria L. D. Palomares and Nicolas Bailly, "Organizing and Disseminating Marine Biodiversity Information: The FishBase and SeaLifeBase Story," in *Ecosystem Approaches to Fisheries: A Global Perspective*, ed. Villy Christensen and Jay Maclean (Cambridge, UK: Cambridge University Press, 2011), 24–46.

5. See: Rainer Froese, "The Science in FishBase," in Christensen and Maclean, *Ecosystem Approaches to Fisheries*, 47–52.

6. Kostantinos I. Stergiou and Athanassios C. Tsikliras, "Scientific Impact of FishBase: A Citation Analysis," in *Fishes in Databases and Ecosystems*, ed. Maria L. D. Palomares, Kostantinos I. Stergiou, and Daniel Pauly, *Fisheries Centre Research Reports* 14, no. 4 (2006): 2–6.

THE BIG LEAGUES

1. Daniel Pauly, *5 Easy Pieces: How Fishing Impacts Marine Ecosystems* (Washington, DC: Island Press, 2010), 1.
2. Pauly, *5 Easy Pieces*, 2.
3. Peter M. Vitousek, Paul R. Ehrlich, Anne H. Ehrlich, and Pamela A. Matson, "Human Appropriation of the Products of Photosynthesis," *BioScience* 36, no. 6 (1986): 368–73.
4. Vitousek et al., "Human Appropriation of the Products of Photosynthesis."
5. John H. Lawton, "What Will You Give Up?" *Oikos* 71, no. 3 (1994): 353–54.
6. Stuart L. Pimm, *The World According to Pimm: A Scientist Audits the Earth* (New York: McGraw-Hill, 2001).
7. T. Radford, "New Calculations by Scientists Show There Can Be No Winners in Fight for Dwindling Catches," *Guardian*, March 16, 1995; M. Ritter, "Global Fishing Taking Up Much of Ocean's Algae, Study Says," Associated Press, March 16, 1995. Quoted in Pauly, *5 Easy Pieces*, 25–27.

8. Daniel Pauly, "Anecdotes and the Shifting Baseline Syndrome of Fisheries," *Trends in Ecology and Evolution* 10, no. 10 (1995): 430.
9. Daniel Pauly, "The Ocean's Shifting Baseline," April 2010, TED Talk, video, 8:47, ted.com/talks/daniel_pauly_the_ocean_s_shifting_baseline.
10. Dietmar Rost, *Wandel (v)erkennen: Shifting Baselines und die Wahrnehmung umweltrelevanter Veränderungen aus wissenssoziologischer Sicht* (Berlin: Springer-Verlag, 2013). [In German.] A short account of his ideas in English may be found in Dietmar Rost, *Shifting Baselines: Interdisciplinary Perspectives on Long-Term Change Perception and Memory*, 2018, nbn-resolving.de/urn:nbn:de:0168-ssoar-56971-0.
11. Kurlansky, *Cod: A Biography*.
12. Pauly, "Anecdotes and the Shifting Baseline Syndrome."
13. Jeremy Jackson, "How We Wrecked the Ocean," April 2010, TED Talk, video, 18:04, ted.com/talks/jeremy_jackson.
14. Jeremy Jackson and Jennifer Jacquet, "The Shifting Baselines Syndrome: Perception, Deception, and the Future of Our Oceans," in Christensen and Maclean, *Ecosystem Approaches to Fisheries*, 128–41.
15. Jeremy B. Jackson, "Reefs Since Columbus," *Coral Reefs* 16, no. 1 (1997): S23–32.
16. Jeremy B. C. Jackson, Michael X. Kirby, Wolfgang H. Berger, Karen A. Bjorndal, Louis W. Botsford,

Bruce J. Bourque, Roger H. Bradbury et al., "Historical Overfishing and the Recent Collapse of Coastal Ecosystems," *Science* 293, no. 5530 (2001): 629–37.

17. Heike K. Lotze and Inka Milewski, "Two Centuries of Multiple Human Impacts and Successive Changes in a North Atlantic Food Web," *Ecological Applications* 14, no. 5 (2004): 1428–47.

18. Heike K. Lotze, "Radical Changes in the Wadden Sea Fauna and Flora Over the Last 2,000 Years," *Helgoland Marine Research* 59, no. 1 (2005): 71–83.

19. Heike K. Lotze, Hunter S. Lenihan, Bruce J. Bourque, Roger H. Bradbury, Richard G. Cooke, Matthew C. Kay, Susan M. Kidwell, Michael X. Kirby, Charles H. Peterson, and Jeremy B. C. Jackson, "Depletion, Degradation, and Recovery Potential of Estuaries and Coastal Seas," *Science* 312, no. 5781 (2006): 1806–9.

20. Heike K. Lotze and Boris Worm, "Historical Baselines for Large Marine Animals," *Trends in Ecology and Evolution* 24, no. 5 (2009): 254–62.

21. Andrea Sáenz-Arroyo, Callum M. Roberts, Jorge Torre, Micheline Cariño-Olvera, and Roberto Ramón Enríquez Andrade, "Rapidly Shifting Environmental Baselines Among Fishers of the Gulf of California," *Proceedings of the Royal Society of London B: Biological Sciences* 272, no. 1575 (2005): 1957–62.

22. Ian Smith, "Retreat and Resilience: Fur Seals and Human Settlement in New Zealand," in *The Exploitation and Cultural Importance of Sea Mammals*, ed. Gregory Monks (Oxford: Oxbow Books, 2005), 6–18.

FISHING DOWN MARINE FOOD WEBS

1. Pauly, *5 Easy Pieces*, 32.

2. Pauly, *5 Easy Pieces*, 32.

3. Daniel Pauly, Villy Christensen, Johanne Dalsgaard, Rainer Froese, and Francisco Torres, "Fishing Down Marine Food Webs," *Science* 279, no. 5352 (1998): 860–63.

4. Pauly et al., "Fishing Down Marine Food Webs,."

5. See also: Konstantinos I. Stergiou and Villy Christensen, "Fishing Down the Food Web," in Christensen and Maclean, *Ecosystem Approaches to Fisheries*, 72–88.

6. Timothy R. Parsons, Masayuki Takahashi, and Barry Hargrave, *Biological Oceanographic Processes* (Oxford: Pergamon Press, 1984).

7. William K. Stevens, "Man Moves Down the Marine Food Chain, Creating Havoc," *New York Times*, February 10, 1998.

8. Stevens, "Man Moves Down the Marine Food Chain."

9. Nancy Baron, "The Scientist As Communicator," in Christensen and Maclean, *Ecosystem Approaches to Fisheries*, 295–303.

10. Baron, "The Scientist As Communicator."

11. Bénédicte Martin, "Comment

scientifiques et écologistes peuvent travailler ensemble. L'exemple de la surpêche," *Reporterre*, September 14, 2015.

12. Pauly, *5 Easy Pieces*, 45.

13. John F. Caddy, Jorge Csirke, Serge Garcia, and Richard Grainger, with response by Daniel Pauly, Rainer Froese, and Villy Christensen, "How Pervasive Is 'Fishing Down Marine Food Webs'?" *Science* 282, no. 5393 (November 20, 1998): 1383.

14. Pauly, *5 Easy Pieces*, 49.

15. Pauly, *5 Easy Pieces*, 56.

16. Trevor A. Branch, Reg Watson, Elizabeth A. Fulton, Simon Jennings, Carey R. McGilliard, Grace T. Pablico, Daniel Ricard, and Sean R. Tracey, "The Trophic Fingerprint of Marine Fisheries," *Nature* 468, no. 7322 (2010): 431–35.

17. Trevor A. Branch, "Fishing Impacts on Food Webs: Multiple Working Hypotheses," *Fisheries* 40, no. 8 (2015): 373–75.

THE SEA AROUND US

1. Joshua Reichert, "Foreword," in *Global Atlas of Marine Fisheries: A Critical Appraisal of Catches and Ecosystem Impacts*, ed. Daniel Pauly and Dirk Zeller (Washington, DC: Island Press, 2016), xi.

2. Villy Christensen, Sylvie Guénette, Johanna J. Heymans, Carl J. Walters, Reginald Watson, Dirk Zeller, and Daniel Pauly, "Hundred-Year Decline of North Atlantic Predatory Fishes," *Fish and Fisheries* 4, no. 1 (2003): 1–24.

3. Jeffrey A. Hutchings, "Ransom Aldrich Myers (1952–2007): In Memoriam," *Canadian Journal of Fisheries and Aquatic Sciences* 65, no. 1 (2008): xii–xix.

4. Hutchings, "Ransom Aldrich Myers."

5. Boris Worm and Ransom A. Myers, "Meta-Analysis of Cod-Shrimp Interactions Reveals Top-Down Control in Oceanic Food Webs," *Ecology* 84, no. 1 (2003): 162–73.

6. Julia K. Baum, Ransom A. Myers, Daniel G. Kehler, Boris Worm, Shelton J. Harley, and Penny A. Doherty, "Collapse and Conservation of Shark Populations in the Northwest Atlantic," *Science* 299, no. 5605 (2003): 389–92.

7. Ransom A. Myers and Boris Worm, "Rapid Worldwide Depletion of Predatory Fish Communities," *Nature* 423, no. 6937 (2003): 280–83.

8. "Les stocks de grands poissons en danger de disparition," *Le Monde*, May 17, 2003.

9. Daniel Pauly, "Obituary: Ransom Aldrich Myers (1952–2007)," *Nature* 447, no. 7141 (2007): 160.

10. Carl Walters, "Folly and Fantasy in the Analysis of Spatial Catch Rate Data," *Canadian Journal of Fisheries and Aquatic Sciences* 60, no. 12 (2003): 1433–36.

11. Pauly, "Obituary: Ransom Aldrich Myers."

12. Daniel Pauly and Jay Maclean, *In a Perfect Ocean: The State of Fisheries and Ecosystems in the North Atlantic Ocean* (Washington, DC: Island Press, 2003), 63.

13. Pauly and Maclean, *In a Perfect Ocean*, 77.

14. Pauly and Maclean, *In a Perfect Ocean*, 90.

15. Pauly and Maclean, *In a Perfect Ocean*, 93.

CHINESE FISHERIES AND CHARLES DARWIN

1. Pauly, *5 Easy Pieces*, 63.

2. Pauly, *5 Easy Pieces*, 63.

3. Pauly, *5 Easy Pieces*, 65.

4. Reg Watson and Daniel Pauly, "Systematic Distortions in World Fisheries Catch Trends," *Nature* 414, no. 6863 (2001): 534–36.

5. Pauly, *5 Easy Pieces*, 90.

6. Pauly, *5 Easy Pieces*, 82.

7. Pauly, *5 Easy Pieces*, 155.

8. Pauly, *5 Easy Pieces*, 89.

9. Daniel Pauly, Villy Christensen, Sylvie Guénette, Tony J. Pitcher, U. Rashid Sumaila, Carl J. Walters, Reg Watson, and Dirk Zeller, "Towards Sustainability in World Fisheries," *Nature* 418, no. 6898 (2002): 689–95.

10. Daniel Pauly, Jackie Alder, Elena Bennett, Villy Christensen, Peter Tyedmers, and Reg Watson, "The Future for Fisheries," *Science* 302, no. 5649 (2003): 1359–61.

11. Jonathan Weiner, *The Beak of the Finch: A Story of Evolution in Our Time* (New York: Vintage, 1994).

12. R. B. Freeman and Charles Darwin, *Works of Charles Darwin* (Norwich, UK: Dawson, 1977).

13. Daniel Pauly, *Darwin's Fishes: An Encyclopedia of Ichthyology, Ecology, and Evolution* (Cambridge, UK: Cambridge University Press, 2004), xiv.

UNCERTAIN GLORY

1. Carol Kaesuk Yoon, "Scientist at Work: Daniel Pauly; Iconoclast Looks for Fish and Finds Disaster," *New York Times*, January 21, 2003.

2. Ray Hilborn, quoted in *Science*, April 19, 2002.

3. Charles Darwin, *The Autobiography of Charles Darwin 1809–1882, With Original Omissions Restored*, ed. Nora Barlow (New York: Harcourt, Brace and Company, 1958).

4. *Nature*, January 2, 2003.

5. G. Bruce Knecht, *Hooked: Pirates, Poaching, and the Perfect Fish* (Emmaus, PA: Rodale, 2007).

6. Kerri Smith, "Love in the Lab: Close Collaborators," *Nature* 510, no. 7506 (2014): 458–60.

7. Christensen and Maclean, *Ecosystem Approaches to Fisheries*.

RECONSTRUCTIONS

1. Kristin Kaschner, "Modelling and Mapping Resource Overlap Between Marine Mammals and Fisheries on a Global Scale"(doctoral thesis, University of British Columbia, 2004).

2. Peter Yodzis, "Must Top Predators Be Culled for the Sake of

Fisheries?" *Trends in Ecology & Evolution* 16, no. 2 (2001): 78–84.

3. Vasiliki S. Karpouzi, Reg Watson, and Daniel Pauly, "Modelling and Mapping Resource Overlap Between Seabirds and Fisheries on a Global Scale: A Preliminary Assessment," *Marine Ecology Progress Series* 343 (2007): 87–99. For an update, see David Grémillet, Aurore Ponchon, Michelle Paleczny, Maria-Lourdes D. Palomares, Vasiliki Karpouzi, and Daniel Pauly, "Persisting Worldwide Seabird-Fishery Competition Despite Seabird Community Decline," *Current Biology* 28, no. 24 (2018), 4009–13.

4. Daniel Pauly and Dirk Zeller, eds., *Global Atlas of Marine Fisheries: A Critical Appraisal of Catches and Ecosystem Impacts* (Washington, DC: Island Press, 2016), 6.

5. Pauly and Zeller, *Global Atlas of Marine Fisheries*, 5.

6. Pauly and Zeller, *Global Atlas of Marine Fisheries*, 8.

7. Daniel Pauly and Dirk Zeller, "Catch Reconstructions Reveal That Global Marine Fisheries Catches Are Higher Than Reported and Declining," *Nature Communications* 7, no. 10244 (2016).

AFRICA FOREVER

1. J. Michael Vakily, Katy Seto, and Daniel Pauly, *The Marine Fisheries Environment of Sierra Leone: Belated Proceedings of a National Seminar Held in Freetown, 25–29 November* 1991, *Fisheries Centre Research Reports* 20, no. 4 (2012).

2. Birane Samb and Daniel Pauly, "On 'Variability' as a Sampling Artefact: The Case of Sardinella in North-Western Africa," *Fish and Fisheries* 1, no. 2 (2000): 206–10.

3. Timothy R. Baumgartner, A. Soutar, and Vicente Ferreira-Bartrina, "Reconstruction of the History of Pacific Sardine and Northern Anchoveta Populations Over the Past Two Millennia From Sediments of the Santa Barbara Basin, California," *CalCOFI Rep* 33 (1992): 24–40.

4. André Fontana and Alassane Samba, eds., *Artisans de la mer: Une histoire de la pêche maritime sénégalaise* (Dakar, Senegal: La Rochette, 2003).

5. Maria Lourdes Palomares and Daniel Pauly, eds., *West African Marine Ecosystems: Models and Fisheries Impacts, Fisheries Centre Research Reports* 12, no. 7 (2004).

6. For the English edition, see: Michel Beuret, Serge Michel, and Paolo Woods, *China Safari: On the Trail of Beijing's Expansion in Africa*, trans. Raymond Valley (New York: Nation Books, 2009).

7. Pauly and Zeller, *Global Atlas of Marine Fisheries*.

8. Daniel Pauly, Dyhia Belhabib, Roland Blomeyer, William W. Cheung, Andrés M. Cisneros-Montemayor, Duncan Copeland, Sarah Harper et al., "China's Distant-Water Fisheries in the 21st

Century," *Fish and Fisheries* 15, no. 3 (2014): 474–88.

9. Daniel Pauly, "Worrying About Whales Instead of Managing Fisheries: A Personal Account of a Meeting in Senegal," *Sea Around Us Project Newsletter* 47, no. 1 (2008).

10. Leah R. Gerber, Lyne Morissette, Kristin Kaschner, and Daniel Pauly, "Should Whales Be Culled to Increase Fishery Yield?" *Science* 323, no. 5916 (2009): 880-81.

11. Christian Chaboud, Massal Fall, Jocelyne Ferraris, André Fontana, Alain Fonteneau, Francis Laloë, Alassane Samba, and Djiga Thiao, "Comment on 'Fisheries Catch Misreporting and Its Implications: The Case of Senegal,'" *Fisheries Research* 164 (2015): 322–24.

12. Dyhia Belhabib, Viviane Koutob, Aliou Sall, Vicky W. Y. Lam, Dirk Zeller, and Daniel Pauly, "Counting Pirogues and Missing the Boat: Reply to Chaboud et al.'s Comment on Belhabib et al. 'Fisheries Catch Misreporting and Its Implications: The Case of Senegal,'" *Fisheries Research* 164 (2015): 325–28.

FRENCH ALLIES

1. Philippe Cury and Yves Miserey, *Une mer sans poissons* (Paris: Calmann-Lévy, 2008), 16.

2. Cury and Miserey, *Une mer sans poissons*, 198.

3. Benoît Mesnil, "Public-Aided Crises in the French Fishing Sector,"

Ocean & Coastal Management 51 (2008): 689–700.

4. Gaëlle Dupont, "Il faut pêcher moins si l'on veut continuer à pouvoir pêcher" (We must fish less if we want to continue to be able to fish), *Le Monde*, March 3, 2009; M. Henry, "Daniel Pauly ne laisse rien filet" (Daniel Pauly lets nothing through the net), *Libération*, March 14, 2009.

5. Telmo Morato, Reg Watson, Tony J. Pitcher, and Daniel Pauly, "Fishing Down the Deep," *Fish and Fisheries* 7, no. 1 (2006): 24–34.

6. Cury and Miserey, *Une mer sans poissons*, 99.

7. "BLOOM—10 ans," fr.calameo. com/read/00468861079 bc9fd08a76, 44.

8. Claire Nouvian, *The Deep: Extraordinary Creatures of the Abyss* (Chicago: University of Chicago Press, 2007).

9. "BLOOM—10 ans," 44.

10. "BLOOM—10 ans," 45.

11. "BLOOM—10 ans," 45.

12. "BLOOM—10 ans," 48.

13. Jennifer Jacquet, *Is Shame Necessary?: New Uses for an Old Tool* (New York: Vintage, 2016).

14. Philippe Cury and Daniel Pauly, *Mange tes méduses! Réconcilier les cycles de la vie et la flèche du temps* (Paris: Odile Jacob, 2013). Published in English (with foreword by Paul Ehrlich) as *Obstinate Nature* (Paris: Odile Jacob, 2021; ebook, hard copy available on request).

332 | THE OCEAN'S WHISTLEBLOWER

15. Jennifer Jacquet and Daniel Pauly, "The Rise of Seafood Awareness Campaigns in an Era of Collapsing Fisheries," *Marine Policy* 31, no. 3 (2007): 308–13.

16. Jennifer Jacquet and Daniel Pauly, "Trade Secrets: Renaming and Mislabeling of Seafood," *Marine Policy* 32, no. 3 (2008): 309–18.

17. Jennifer Jacquet, Daniel Pauly, David Ainley, Sidney Holt, Paul Dayton, and Jeremy Jackson, "Seafood Stewardship in Crisis," *Nature* 467, no. 7311 (2010): 28–29.

18. Cury and Pauly, *Mange tes méduses!*, 181.

19. Cury and Pauly, *Mange tes méduses!*, 13.

20. Cury and Pauly, *Mange tes méduses!*, 17.

21. Cury and Pauly, *Mange tes méduses!*, 71.

22. Cury and Pauly, *Obstinate Nature*, chapter 2.

23. Cury and Pauly, *Mange tes méduses!*, 123.

24. See also: Paul Virilio, *Speed and Politics* (Cambridge, MA: MIT Press, 2007).

FIRST LOVES, FINAL BATTLES

1. Daniel Pauly, *Gasping Fish and Panting Squids: Oxygen, Temperature and the Growth of Water-Breathing Animals* (Oldendorf/Luhe, Germany: International Ecology Institute, 2010), 144.

2. William W. Cheung and Yvonne Sadovy, "Retrospective Evaluation of Data-Limited Fisheries: A Case From Hong Kong," *Reviews in Fish Biology and Fisheries* 14, no. 2 (2004): 181–206.

3. William W. L. Cheung, Vicky W. Y. Lam, Jorge L. Sarmiento, Kelly Kearney, Reg Watson, and Daniel Pauly, "Projecting Global Marine Biodiversity Impacts Under Climate Change Scenarios," *Fish and Fisheries* 10, no. 3 (2009): 235–51.

4. William W. L. Cheung, Vicky W. Y. Lam, Jorge L. Sarmiento, Kelly Kearney, Reg Watson, Dirk Zeller, and Daniel Pauly, "Large-Scale Redistribution of Maximum Fisheries Catch Potential in the Global Ocean Under Climate Change," *Global Change Biology* 16, no. 1 (2010): 24–35.

5. William W. L. Cheung, Reg Watson, and Daniel Pauly, "Signature of Ocean Warming in Global Fisheries Catch," *Nature* 497, no. 7449 (2013): 365–68.

6. William W. L. Cheung, Jorge L. Sarmiento, John Dunne, Thomas L. Frölicher, Vicky W. Y. Lam, M. L. Deng Palomares, Reg Watson, and Daniel Pauly, "Shrinking of Fishes Exacerbates Impacts of Global Ocean Changes on Marine Ecosystems," *Nature Climate Change* 3, no. 3 (2013): 254–58.

7. Sjannie Lefevre, David J. McKenzie, and Göran E. Nilsson, "Models Projecting the Fate of Fish Populations Under Climate Change Need to Be Based on Valid

Physiological Mechanisms," *Global Change Biology* 23 (2017): 3449–59.

8. Daniel Pauly and William W. L. Cheung, "Sound Physiological Knowledge and Principles in Modeling Shrinking of Fishes Under Climate Change," *Global Change Biology* 24, no. 1 (2018): e15–e26.

9. Sjannie Lefevre, David J. McKenzie, and Göran E. Nilsson, "In Modelling Effects of Global Warming, Invalid Assumptions Lead to Unrealistic Projections," *Global Change Biology* 24 (2018): 553–56.

10. Ray Hilborn, "Current and Future Trends in Fisheries Stock Assessment and Management," *South African Journal of Marine Science* 12, no. 1 (1992): 975–88.

11. Peter Calamai, "The Fisher King," *Cosmos Magazine* (Australia), December 2006.

12. Boris Worm, Edward B. Barbier, Nicola Beaumont, J. Emmett Duffy, Carl Folke, Benjamin S. Halpern, Jeremy B. C. Jackson et al., "Impacts of Biodiversity Loss on Ocean Ecosystem Services," *Science* 314, no. 5800 (2006): 787–90.

13. Boris Worm, Ray Hilborn, Julia K. Baum, Trevor A. Branch, Jeremy S. Collie, Christopher Costello, Michael J. Fogarty et al., "Rebuilding Global Fisheries," *Science* 325, no. 5940 (2009): 578–85.

14. Hilborn, "Current and Future Trends in Fisheries Stock Assessment and Management."

15. Stéphane Foucart and Martine Valo, "Ray Hilborn, un expert de la pêche dans la tempête," *Le Monde*, May 14, 2016, 7.

16. Foucart and Valo, "Ray Hilborn, un expert de la pêche."

SELECTED BIBLIOGRAPHY

Cheung, William W., Vicky W. Lam, Jorge L. Sarmiento, Kelly Kearney, Reg Watson, and Daniel Pauly. "Projecting Global Marine Biodiversity Impacts Under Climate Change Scenarios." *Fish and Fisheries* 10, no. 3 (2009): 235–51.

Cheung, William W., Vicky W. Lam, Jorge L. Sarmiento, Kelly Kearney, Reg Watson, Dirk Zeller, and Daniel Pauly. "Large-Scale Redistribution of Maximum Fisheries Catch Potential in the Global Ocean Under Climate Change." *Global Change Biology* 16, no. 1 (2010): 24–35.

Cheung, William W., Jorge L. Sarmiento, John Dunne, Thomas L. Frölicher, Vicky W. Y. Lam, M. L. Deng Palomares, Reg Watson, and Daniel Pauly. "Shrinking of Fishes Exacerbates Impacts of Global Ocean Changes on Marine Ecosystems." *Nature Climate Change* 3, no. 3 (2013): 254–58.

Cheung, William W., Reg Watson, and Daniel Pauly. "Signature of Ocean Warming in Global Fisheries Catch." *Nature* 497, no. 7449 (2013): 365–68.

Christensen, Villy, Sylvie Guénette, Johanna J. Heymans, Carl J. Walters, Reg Watson, Dirk Zeller, and Daniel Pauly. "Hundred-Year Decline of North Atlantic Predatory Fishes." *Fish and Fisheries* 4, no. 1 (2003): 1–24.

Christensen, Villy, and Daniel Pauly. "ECOPATH II–A Software for Balancing Steady-State Ecosystem Models and Calculating Network Characteristics." *Ecological Modelling* 61, no. 34 (1992): 169–85.

Christensen, Villy, and Daniel Pauly, eds. *Trophic Models of Aquatic Ecosystems.* ICLARM Conference Proceedings 26. Makati City, Philippines: ICLARM, 1993.

Cury, Philippe, and Daniel Pauly. *Mange tes méduses! Réconcilier les cycles de la vie et la flèche du temps.* Paris: Odile Jacob, 2013. [In French.]

Fréon, Pierre, Marilú Bouchon, Christian Mullon, Christian García, and Miguel Ñiquen. "Interdecadal Variability of Anchoveta Abundance and Overcapacity of the Fishery in Peru." *Progress in Oceanography* 79, no. 2–4

(October–December 2008): 401–12.

Froese, Rainer, and Daniel Pauly, eds. *FishBase 2000: Concepts, Design and Data Sources*. Makati City, Philippines: ICLARM, 2000.

Gerber, Leah R., Lyne Morissette, Kristin Kaschner, and Daniel Pauly. "Should Whales Be Culled to Increase Fishery Yield?" *Science* 323, no. 5916 (2009): 880–81.

Jacquet, Jennifer L., and Daniel Pauly. "Trade Secrets: Renaming and Mislabeling of Seafood." *Marine Policy* 32, no. 3 (2008): 309–18.

Jacquet, Jennifer, Daniel Pauly, David Ainley, Sidney Holt, Paul Dayton, and Jeremy Jackson. "Seafood Stewardship in Crisis." *Nature* 467, no. 7311 (2010): 28–29.

Longhurst, Alan R., and Daniel Pauly. *Ecology of Tropical Oceans*. Cambridge, MA: Academic Press, 1987.

Morato, Telmo, Reg Watson, Tony J. Pitcher, and Daniel Pauly. "Fishing Down the Deep." *Fish and Fisheries* 7, no. 1 (2006): 24–34.

Palomares, Maria L., and Daniel Pauly, eds. *West African Marine Ecosystems: Models and Fisheries Impacts*. Fisheries Centre Research Reports 12, no. 7 (2004).

Palomares, Maria L. D., Konstantinos I. Stergiou, and Daniel Pauly, eds. *Fishes in Databases and Ecosystems*. Fisheries Centre Research Reports 14, no. 4 (2006): 2–6.

Pauly, Daniel. "Anecdotes and the Shifting Baseline Syndrome of Fisheries." *Trends in Ecology and Evolution* 10, no. 10 (1995): 430.

Pauly, Daniel. "Bericht über die amerikanische Welszuchtindustrie" [Report on the U.S. catfish industry]. In *Neue Erkenntnisse auf dem Gebiet der Aquakultur*, vol. 16, *Arbeiten des Deutschen Fischereiverbandes*, edited by K. Tiews, 154–67. Hamburg: Deutscher Fischerei-Verband, 1974. [In German.]

Pauly, Daniel. *Darwin's Fishes: An Encyclopedia of Ichthyology, Ecology, and Evolution*. Cambridge, UK: Cambridge University Press, 2004.

Pauly, Daniel. "A Device for Presorting Benthos Samples." *Berichte der Deutschen wissenschaftlichen Kommission für Meeresforschung* 22, no. 4 (1973): 458–60. [In German with an abstract in English and a complete translation of the text.]

Pauly, Daniel. *5 Easy Pieces: How Fishing Impacts Marine Ecosystems*. Washington, DC: Island Press, 2010.

Pauly, Daniel. *Gasping Fish and Panting Squids: Oxygen, Temperature and the Growth of Water-Breathing Animals*. Oldendorf/Luhe, Germany: International Ecology Institute, 2010.

Pauly, Daniel. "The Gill-Oxygen Limitation Theory (GOLT) and Its Critics." *Science Advances* 7 (2021): eabc6050.

Pauly, Daniel. "Gill Size and Temperature as Governing Factors in Fish Growth: A Generalization of von Bertalanffy's Growth Formula." Doctoral thesis, Institut für Meereskunde, Kiel University, 1979.

Pauly, Daniel. "Major Trends in Small-Scale Marine Fisheries, With Emphasis on Developing Countries, and Some Implications for the Social Sciences." MAST 4, no. 2 (2006): 7–22.

Pauly, Daniel. *Méthodes pour l'évaluation des ressources halieutiques.* Translated by J. Moreau. Toulouse: Collection Polytech de l'INP de Toulouse, 1997. [In French.]

Pauly, Daniel. "On the Ecology of a Small West African Lagoon." *Berichte des Deutschen wissenschaftlichen Kommission für Meeresforschung* 24, no. 1 (1975): 46–62.

Pauly, Daniel. "On the Interrelationships Between Natural Mortality, Growth Parameters, and Mean Environmental Temperature in 175 Fish Stocks." *ICES Journal of Marine Science* 39, no. 2 (1980): 175–92.

Pauly, Daniel. *On the Sex of Fish and the Gender of Scientists: A Collection of Essays in Fisheries Science.* Fish and Fisheries Series, vol. 6. London: Chapman & Hall, 1994.

Pauly, Daniel. "Some Definitions of Overfishing Relevant to Coastal Zone Management in Southeast Asia." *Tropical Coastal Area Management* 3, no. 1 (1988): 14–15.

Pauly, Daniel. "Theory and Management of Tropical Multispecies Stocks: A Review, With Emphasis on the Southeast Asian Demersal Fisheries." ICLARM Studies and Reviews 1 (1979).

Pauly, Daniel. "Why Squid, Though Not Fish, May Be Better Understood by Pretending They Are." *South African Journal of Marine Science* 20, no. 1 (1998): 47–58.

Pauly, Daniel, Jackie Alder, Elena Bennett, Villy Christensen, Peter Tyedmers, and Reg Watson. "The Future for Fisheries." *Science* 302, no. 5649 (2003): 1359–61.

Pauly, Daniel, Dyhia Belhabib, Roland Blomeyer, William W. Cheung, Andrés M. Cisneros-Montemayor, Duncan Copeland, Sarah Harper et al. "China's Distant-Water Fisheries in the 21st Century." *Fish and Fisheries* 15, no. 3 (2014): 474–88.

Pauly, Daniel, and William W. Cheung. "Sound Physiological Knowledge and Principles in Modeling Shrinking of Fishes Under Climate Change." *Global Change Biology* 24, no. 1 (2018): e15–e26.

Pauly, Daniel, and Villy Christensen. "Primary Production Required to Sustain Global Fisheries." *Nature* 374 (1995): 255–57.

Pauly, Daniel, Villy Christensen, Johanne Dalsgaard, Rainer Froese, and Francisco Torres. "Fishing Down Marine Food Webs." *Science* 279, no. 5352 (1998): 860–63.

Pauly, Daniel, Villy Christensen, Sylvie Guénette, and Tony J. Pitcher. "Towards Sustainability in World Fisheries." *Nature* 418, no. 6898 (2002): 689–95.

Pauly, Daniel, and Rainer Froese. "FishBase: Assembling Information on Fish." *Naga, the ICLARM Quarterly* 9, no. 4 (1991): 10–11.

Pauly, Daniel, and Jay Maclean. *In a Perfect Ocean: The State of Fisheries and Ecosystems in the North Atlantic Ocean*. Washington, DC: Island Press, 2003.

Pauly, Daniel, Peter Muck, Jaime Mendo, and Isabel Tsukayama, eds. *The Peruvian Upwelling Ecosystem: Dynamics and Interactions*. Callao, Peru: Instituto del Mar del Perú, 1989.

Pauly, Daniel, and Isabel Tsukayama. "The Peruvian Anchoveta and Upwelling Ecosystem: Three Decades of Change." ICLARM Studies and Reviews 15 (1987).

Pauly, Daniel, and Dirk Zeller. "Catch Reconstructions Reveal That Global Marine Fisheries Catches Are Higher Than Reported and Declining." *Nature Communications* 7 (2016).

Pauly, Daniel, and Dirk Zeller, eds. *Global Atlas of Marine Fisheries: A Critical Appraisal of Catches and Ecosystem Impacts*. Washington, DC: Island Press, 2016.

Venema, Siebren C., Jörgen M. Christensen, and Daniel Pauly. "Training in Tropical Fish Stock Assessment: A Narrative of Experience." In *Contributions to Tropical Fisheries Biology: Papers by the Participants of* FAO/DANIDA *Follow-Up Training Courses*, edited by Siebren Venema, Jörgen Möller-Christensen, and Daniel Pauly, 1–15. FAO Fisheries Report 389 (1988).

Walters, Carl, Villy Christensen, and Daniel Pauly. "Structuring Dynamic Models of Exploited Ecosystems From Trophic Mass-Balance Assessments." *Reviews in Fish Biology and Fisheries* 7, no. 2 (1997): 139–72.

Watson, Reg, and Daniel Pauly. "Systematic Distortions in World Fisheries Catch Trends." *Nature* 414, no. 6863 (2001): 534–36.

ACKNOWLEDGMENTS

I WOULD LIKE TO thank Daniel Pauly for placing so much trust in me and for three years of intense conversations. Thanks are also owed to the Wade/Pauly families for tolerating my intrusions with friendly benevolence, to all the people mentioned in this book, as well as to many others whose interviews I unfortunately could not include. My most sincere thanks to Suzy and Jef Blaser for their research in Switzerland, to Stephane de Giorgi for finding the names of Daniel's schoolteachers, to Gerrit Peters and Ute Meischner (Kiel), Margie and Jay Maclean (Melon Patch), Deng Palomares and Nicolas Bailly (Manila), Heather Maclean (Manila), James Lamar Gibson and Roodline Volcy (Greensboro), Mélanie Guigueno and Kyle Elliott (Montreal), Jacob González-Solis (Barcelona), Serigne Diagne (Dakar), Eva and Cyril Peysson (Paris), Claire Nouvian (Paris), and the Lauret-Grémillet family (Paris) for their hospitality, and Veronique Garçon for looking over the chapter on Peru. Thanks also to Bénédicte Martin, Sophie Grémillet, Léna Widerkehr, and Daniel Cheville for their attentive readings of the entire manuscript, as well as all the people who encouraged me to take up writing: the whole Grémillet family, but also David Quammen, Olivier Dangles, Amélie Lescroël, Raphaël Mathevet, and Jacques Blondel. Thanks are also owed to Athanassios Tsikliras, Michel Gauthier-Clerc, Donna Dimarchopoulou, Myriam Khalfallah, and Nazanine Hozar for their support. I've been immensely fortunate to meet my French editor, Baptiste Lanaspeze of Éditions Wildproject, and the translator of this epos, Georgia Lyon Froman—I thank them both warmly for their liberating enthusiasm and their attention to detail whilst working on the manuscript. From there, our distinguished Canadian colleagues from Greystone Books took over, and I thank them for their highly professional and friendly attitude all through the preparation of the final English edition. Thanks especially to Paula Ayer, who edited the translated manuscript, designers Fiona Siu and Jessica Sullivan, and proofreader Jennifer Stewart. And finally, thank you Bénédicte and Ambalika, for always being there for me in this life of thrilling, but sometimes solitary, labor.

This work has been partly funded by the Oak Foundation, with much support from Kristian Parker, but the views expressed are those of the author alone.

PHOTO CREDITS

THE IMAGES IN the photo album in the middle of the book are from the collection of the Pauly family, except for the following: Wade family portrait with Daniel in Seaside, California, 1980 (personal collection of Dr. Cheryl Wade); Daniel going diving (© Roger Pullin); Cousin Ed Whitfield in 2016 in Greensboro, Jennifer Jacquet, Cornelia Nauen and Gotthilf Hempel, Daniel Pauly and Philippe Cury (© David Grémillet); all of the photographs of Deng Palomares (personal collection of Deng Palomares); Daniel with Walter Kühhirt (personal collection of Walter Kühhirt); Daniel with Jay Maclean, diving in Tubbataha (personal collection of Jay Maclean); Daniel receiving the Cosmos Prize (© Cosmos Prize); Senegalese fishermen (© Anna Badkhen).

INDEX

Photographs indicated by page numbers in italics

Acosta, Belen, 153
activism: ecolabeling, 293–94, 306; impact
    of environmental NGOs, 314; limited
    power of consumer activism, 294–95;
    Pauly's relationship with NGOs, 212, 227;
    research and, 274
Adenauer, Konrad, 19
Africa, 263, 282–83. *See also* Ghana; Senegal;
    Sierra Leone; West Africa
African Americans, 2–3, 12, 33–36, 80. *See
    also* Black people, in Europe
Albert I medal, *177*, 290
anchoveta, 121, 122, 123–27, 128, 244
Andrea Gail (fishing vessel), 235
aquaculture, 57, 60, 65, 103–4, 115, 214
Arntz, Wolf "Petz," 51–52, 53, 81, 82, 122–23,
    124, 299
Aronson, Lester, 60
artisanal fisheries: development challenges
    in West Africa, 278, 279; importance of,
    85, 239, 278; Sakumo Lagoon (Ghana),
    56–57; San Miguel Bay (Philippines), 92;
    Senegal, *179*; Sierra Leone, 268
ascendency, 143
Atanacio, Rachel "Aque," *176*, 210–11, 212
Atkinson, Marlin, 138

Bailly, Nicolas, 156–57
Baranov, Fedor, 59
Barbier, Jacques, 14, 15–17
Barnier, Michel, 291
Baron, Nancy, 209–10, 211, 242, 254

Baumgartner, Timothy R., 270
Bayesian statistics, 100–101
Bay of Fundy, 202–3
Beddington, John, 194
Belhabib, Dyhia, 277–79, 280–81
Berg, Lev, 59
Bertrand, Arnaud, 121, 126, 127
Beverton, Raymond, 73, 75
Bianchi, Gabriella, 270, 271
Binohlan, Cris, 155, 156
BirdLife International, 266
birds, marine, 260, 266–67
Black people, in Europe, 19–20, 80. *See also*
    African Americans
Blaser, Jean-François "Jef," 13–14, 15
Blaser, Suzy (née Mauerhofer), 13–14, *14–15*,
    17–18
BLOOM, 290–91
Blue Marine Foundation: Ocean Award, 262
Borloo, Jean-Louis, 287
Branch, Trevor, 215–17
British Virgin Islands, 140
bycatch, 193, 197

Cabanban, Annadel, 117, 258
Caddy, John, 111
California, 121
Canary Current system, 264
Capuli, Emily, 155, 156
Carson, Rachel, 113, 225–27, 256; *The Sea
    Around Us*, 225; *Silent Spring*, 226
Catalonia, 11

catfish, 60
Chavance, Pierre, 271
Cheung, William, 300–303, 309, 313
China, 240–43, 266–67, 275–76, 278, 313
Christensen, Villy: arrival and departure
    from ICLARM, 142, 189; background, 142;
    Ecopath and Ecosim, 142, 143, 144–45,
    148, 172; "Fishing Down Marine Food
    Webs" and, 207, 213; global study of
    fisheries, 192–94, 195; North Atlantic
    study, 229; Pauly's sixtieth birthday
    celebrations and, 257–58; Sea Around Us
    and, 228; SIAP project on West African
    fisheries, 271; at UBC, 257
Clément, Henri (Pauly's grandfather), 1–2,
    29
climate change, 158, 212, 301–3, 312
Close, Chris, 301
Club of Rome: The Limits of Growth, 192
cod and cod fisheries, 47–48, 70, 200, 209,
    230, 231
Coll, Marta, 148
Consultative Group for International
    Agricultural Research (CGIAR), 186–88,
    192
Conti, Anita, 263
Convention on Biological Diversity, 215
coral ecosystems, 137–40, 200–201
Coriolis force, 120
Corpus, Loida, 159
Cousteau, Jacques-Yves, 256
Cruz-Trinidad, Annabelle, 116
Csirke, Jorge, 111
Cury, Philippe: about, 272, 284; on fisheries
    management, 289; on Pauly, 255, 282,
    298, 309; relationship with Pauly, 177,
    285–86; on trawlers, 290; works: Mange
    tes méduses! (Eat your jellyfish!; with
    Pauly), 295–97; Une mer sans poissons (A
    sea without fish; with Miserey), 284–85
Cushing, David, 84
Cyrulnik, Boris, 8

Daget, Jacques, 66
Dalsgaard, Johanne, 207
Damanaki, Maria, 291

Darwin, Charles, 243–46, 247
data: big data, definition, 154n; Dr. Fridtjof
    Nansen data sharing issues, 270–71;
    transparency and open access, 75–76, 124,
    153–54, 273–74
David, Noel, 101
DDT, 226–27
degrowth, 297
Department of Transportation and
    Infrastructure (DDE; France), 28
Diamond, Jared, 77, 255; Collapse, 209
dictionnaire amoureux, 245
Dietrich, Günter, 43
diving, 117–18, 174, 183, 222
Dizon, Leticia, 115–16
Dr. Fridtjof Nansen (research vessel), 269,
    270–71

ecolabeling, 293–94, 306
Ecopath and Ecopath II: ascendency and,
    143; attempts to make dynamic, 146–48;
    development by Polovina, 137–39, 172;
    freshwater ecosystems and, 146; ICES
    conference on, 144–45, 172; Pauly's
    first use of, 139–40; reprogramming by
    Christensen and Pauly, 142–44; trophic
    levels and, 142–43
Ecosim, 147–48
EcoTroph, 274–75
Ehrlich, Paul and Anne: The Population Bomb,
    192
el Ali, Haïdar, 280
ELEFAN (Electronic Length Frequency
    Analysis), 101, 102–3, 109–11, 123, 132,
    139, 304
El Niño–Southern Oscillation (ENSO), 122
Encyclopedia of Life, 158
environmental non-governmental
    organizations (NGOS), 212, 227, 314
Eschmeyer, Bill, 157
European Commission, 45, 153, 159,
    238

Fischer, Walter, 152
fish: age, determining from otoliths, 69–70;
    climate change and, 301–3; gill-oxygen

limitation theory, 73, 76–78, 246, 298–
300, 303; growth of tropical fish, 132;
growth rates, determining, 98–101,
102–3; natural mortality rates (M),
determining, 74–76; trophic levels, 142–
43, 193. *See also* fisheries
FishBase: author's discovery of, 149–50;
critiques of, 157–58; current offices
and status, 154–55, 190; development
of, 151–54; French version, 157; funding
challenges, 159; illustrations, 210–11;
SeaLifeBase and, 158–59; species
classification system, 156, 157; staff, 155–
56, 159, 190
fisheries: anchoveta study and management
in Peru, 123–27, 128; Angela Pauly on
overfishing, 220; China, 240–43, 266–67,
275–76, 278, 313; cod, 47–48, 200, 209,
230, 231; countering overfishing, 232–33,
313–14; ecosystem-based management,
238–39; fishing down trophic levels, 176,
206–10, 211–17, 231; fish population
studies, development of, 47–48; France,
287–89, 312; Germany, 53–54; as
global disaster, 196–97, 312–13; global
exploitation study, 192–96; history of
and historical exploitation, 47, 200–205;
infinite resources myth, 194, 236, 240;
longline fishing, 231–32; Malthusian
overfishing, 94–96; marine mammals,
seabirds, and, 259–60; natural variations
fallacy, 269–70; North Atlantic study,
228–29; Norway, 47–48, 312; *In a Perfect
Ocean* (Pauly and Maclean) on, 233,
235–36, 237–39; pirate fishing, 266;
reconstruction of global catches, 260–62;
reduction fisheries (fish meal plants),
121, 266, 267, 281, 294, 313; Sakumo
Lagoon (Ghana), 56–57, 58–59; *Sea of
Slaughter* (Mowat) on, 199–200; Senegal,
179, 264–65; shifting baseline syndrome,
197–98, 201–2, 205; slavery and, 311n;
social dimensions of, 90–94, 96–97;
traditional management strategies, 48,
69, 237–38; trawlers, 61–62, 66, 92–93,
289–90, 290–91; tropical fisheries, 69,

84–85; uncontrolled fisheries, 266. *See
also* overfishing; West Africa
fisheries biology, 47–49
fish meal plants (reduction fisheries), 121,
266, 267, 281, 294, 313
Fontana, André, 270
Food and Agriculture Organization (FAO):
Chinese catch inflation controversy, 240–
43; critiques of "Fishing Down Marine
Food Webs," 213–15; FishBase and,
152–53; global catches reconstruction
and, 260–62; Nauen and, 45; network of
fish stock management experts, 109–10;
North Atlantic fisheries data, 228; Pauly
and, 129; tropical fisheries and, 85
France: bottom trawling debate, 290–91;
ecologists and fisheries scientists,
263–64, 272–73; fisheries subsidies
and mismanagement, 287–89, 312;
mandatory military service experience,
25–26, 27–28, 66; Pauly and French
fisheries, 286–87, 288–89; Pauly's
relationship with, 234–35; Second World
War and, 1; student protests, 30
France filière pêche, 307
French Frigate Shoals (Kānemiloha'i), 136–37
Fréon, Pierre, 127
Froese, Rainer: arrival and departure from
ICLARM, 153, 188, 189; background, 151–
52; colleagues on, 154, 156; FishBase and,
152–53, 158, 159; "Fishing Down Marine
Food Webs" and, 207, 213; ICLARM Band,
190; on meeting Pauly, 150–51; on Pauly's
children, 217; Pauly's stroke and, 252
Froese, Tom, 154

Gambia, 279
Garcia, Serge, 111
Gascuel, Didier, 272–75, 276, 280, 281, 284,
312, 314
general systems theory, 72
Gerber, Leah, 276–77
German Agency for Technical Cooperation
(GTZ), 65–67, 71, 103
Germany, 1, 19–20, 30, 53–54, 82. *See also*
Kiel; Pauly, Daniel—youth

Ghana, 55–59, 282
glacial erratics, 61–62
Global Change Biology (journal), 303
Grainger, Richard, 240–41, 242
Grandjean, Eva, 8
grape harvesting, 12–13
Greenland, 63–64
Greenpeace, 212, 306
Grenelle de la Mer, 286, 290
Grigg, Richard, 138
grouper, white (*thiof*), 264–65, 267
growth: determining growth rates for fish,
    98–101, 102–3; gill-oxygen limitation
    theory, 73, 76–78, 246, 298–300, 303;
    of tropical fish, 132; von Bertalanffy on,
    72–73
Guardian (newspaper), 195–96
Gulf of Thailand, 66, 84, 85
Gulland, John, 103, 111, 122

Heald, Eric, 143
Helms, Antje, 149–50
Hempel, Gotthilf: on Africa, 282;
    background, 49; Pauly's sixtieth birthday
    celebrations and, 258; and Pauly's
    studies and career, 50–51, 53, 55, 64, 66,
    71–72, 75, 131; photograph, *176*; political
    connections, 141
Henry, Michel, 289
Hensen, Victor, 43
Herberts, Kurt (Herberts Lacke), 22, 26, 29
Hilborn, Ray, 76, 147, 303–7, 312
Hjort, Johan, 48
Holt, Sidney, 73
Hong Kong, 300
Humboldt Current, 122
Hutchings, Jeffrey, 230

Indian Ocean tsunami (2004), 96–97
Indonesia: milkfish fishery, 65; Pauly's
    research with GTZ in, 65–67, 68–71, *165*;
    political tensions, 67–68
Ingles, José "Jingles," 89, 90, 91, 101, 102,
    103–4, 109, 258
Institut de recherche pour le développement
    (IRD), 58, 129, 263, 280–81

Institut français de recherche pour
    l'exploitation de la mer (IFREMER), 286–
    87, 291
Institut für Meereskunde (IfM,
    Oceanographic Institute): about, 43–44;
    contemporary focus, 54; fisheries biology,
    49; Pauly's habilitation to direct research
    at, 129–31, *171*; Pauly's studies at, 42–43,
    44, 46–47, 48, 50–53, 59, 74; women at,
    141
Instituto del Mar del Perú (IMARPE), 121,
    123–26, 128–29
International Center for Living Aquatic
    Resources Management (ICLARM):
    background, 83–84; culture, 114–15;
    decline and departures from, 186–90;
    financial challenges, 129; ICLARM Band,
    190; Pauly's recruitment and departure,
    84, 86, *167*, 188–89; publishing program,
    115; San Miguel Bay project on social
    dimensions of fisheries, 90–94; Sierra
    Leone and, 268; staff recruitments, 105,
    114, 142. *See also* ELEFAN; FishBase
International Cosmos Prize, *183*, 255–57
International Council for the Exploration of
    the Sea (ICES), 75, 144–45, 207
International Ecology Institute, 298, 299
International Rice Research Institute (IRRI),
    154–55, 192
International Whaling Commission, 276
Ivlev, Viktor Sergeevich, 59

Jackson, Jeremy, 200–202, 204, 254
Jacquet, Jennifer, *178*, 292–95, 309, 310, 311
Jamaica, 200–201
Japan, 231–32, 255–57, 276
Jarre, Astrid, 146, 147
jellyfish, 211–12
Jensen, Sabine, 31

Kaltenbach, Marie (Pauly's grandmother),
    1–2
Kambey, Elizabeth, 68, 71, 78
Kanneworff, Per, 62–63
Karibuhoye, Charlotte, 279
Karpouzi, Vasiliki, 260

Kaschner, Kristin, 259
Kiel (Germany), 42–43, 44, 45–46, 53–54. *See also* Institut für Meereskunde
Kinne, Otto, 299
Kleisner, Kristin, 216
Knecht, Bruce, 251
Knowlton, Nancy, 255
Kühhirt, Walter, 23–25, 29, 31, 37, 38, 39–40, 164, 299

La Chaux-de-Fonds (Switzerland), 5. *See also* Pauly, Daniel—childhood
La Fontaine, Jean de, 58
Laloë, Francis, 280
Lam, Vicky, 301
La Pérouse, Jean-François de, 137
Lasker, Reuben, 124
Lefevre, Sjannie, 303
Le Manach, Frédéric, 294
Le Monde (newspaper), 232, 288, 306–7
Libération (newspaper), 288, 289
Liberia, 279
Liberty Times (newspaper), 243
Lightfoot, Clive, 187
Lindeman, Raymond, 142
Little Rock (AR), 33–34
Longhurst, Alan, 131–32, 139
longline fishing, 231–32
Lotze, Heike, 202–5, 254
Lüdi, Lorette, 8
Luna, Susan, 153, 156, 159

Maclean, Jay: arrival and departure from ICLARM, 114, 189; background, 112–14; on Ecopath, 139; ICLARM Band, 190; and merger between ICLARM and CGIAR, 186, 187–88; on Pauly, 115, 116, 117, 118, 133, 136; Pauly's sixtieth birthday celebrations and, 258; Pauly's stroke and, 252; on People Power Revolution in Philippines, 135; *In a Perfect Ocean* (with Pauly), 233, 235–36, 237–39; Pongase persona, 144, 172; publishing program at ICLARM, 115; on Vancouver, 189
Maclean, Margie, 116–17, 135
Majluf, Patricia, 127–28, 141

Malthusian overfishing, 94–96
Manel, Camille, 280
Mansour, Hassan, 216
Maori, 205
marine mammals, 259–60
Marine Stewardship Council (MSC) label, 293–94, 306
marine trophic index, 215
Marr, Jack, 83–84, 85, 86
Martosubroto, Purwito, 67, 68, 71
Matson, Pamela, 192
Mauritania, 267, 274, 279
MAVA, 279
May, Robert "Bob," 153, 194, 196
McCall, Robert, 153
McKenzie, David, 303
McLemore, Carl (Pauly's uncle), 37–38, 60–61
McLemore, Winston R. (Pauly's father), 3, 11, 38–39, 82, 175
McManus, John, 93, 109
media, 195–96, 208–10, 242, 248, 273, 288–89
Mendo, Jaime, 125, 126, 127, 128, 129
Mesnil, Benoît, 287–88
milkfish, 65
Mines, Antonio, 90, 91, 101
Moiseev, Peter, 59
Moloney, Coleen, 141, 246, 282
Monod, Théodore, 263
Morato, Telmo, 289
Moreau, Jacques, 145–46, 284
Morocco, 279
Mowat, Farley, 199, 203; *Sea of Slaughter*, 199–200, 233
Muck, Peter, 126
Mundus maris, 45
Munro, John, 105
Murphy, Garth, 122
Mutiara 4 (research trawler), 68–69, 165
Myers, Ransom, 229–30, 231–33, 254, 305

Namibia (South West Africa), 23–24
National Center for Ecological Analysis and Synthesis, 201–2

Nature (journal), 153, 194–95, 213, 215, 232, 241–42, 243, 248, 294

Nauen, Cornelia: background, 44–45; FishBase and, 153, 159; Hempel and, 49–51; Opitz and, 140; Pauly and, 45, 50, 53, 61, 69, 87, 258, 309; photograph, *176*

Nelson, Joseph, 246

New York Times, 208–9, 242, 248

Nilsson, Göran, 303

Nkrumah, Kwame, 55

Nobel Prize, 256

non-governmental organizations (NGOs), environmental, 212, 227, 314

North Atlantic fisheries, 228–29

Norway, 47–48, 312

Nouvian, Claire, 287, 290–91

Oak Foundation, 158

Obama, Barack, 31n

Oceana, 219, 313

Oceanographic Research Center of Dakar-Thiaroye (CRODT), 267, 280–81

oceanography, 43, 47n, 122–23. *See also* fisheries biology; Institut für Meereskunde

Odum, William, 143

Office de la recherche scientifique et technique outre-mer (ORSTOM), 57–58, 66, 263

Opitz, Silvia, 139–41

otoliths, 69–70

overfishing, 48. *See also* fisheries

Palomares, Maria Lourdes "Deng": anchoveta fishery in Peru and, 123–24, 126; arrival and departure from ICLARM, 105, 189–90; background, 105–9; ELEFAN and, 109–10, 111; FishBase and, 154, 158, 159; marriage, 156–57; Pauly and, 309; photographs, *170*; political activism, 108–9, 135, 136; Sea Around Us funding and, 310, 312; SIAP project and, 271; studies in France, 145–46

Pamulaklakin, Dennis, 90

Pang, Pei, 241

Pantillon, Pierre, 8

Parsons, Timothy, 208

Patagonian toothfish, 251

Paul G. Allen Family Foundation, 310

Pauly, Angela (daughter), 87–88, 133, 141, *169*, *180*, 217–20, 221, 250–51, 311

Pauly, Anita (sister), 30

Pauly, Christian (brother), 29

Pauly, Daniel: about, 314; absentmindedness, 117–18, 133–34; Canadian citizenship, 308; children and parenting, 78, 87, 217–24; diving, 117–18, *174*, *183*, 222; family background, 1–2, 3, 33, *181*; in France, 234–35; impressions of, 79, 98, 115, 116, 118, 150–51, 155, 247, 281, 285, 304; life in Philippines, 87–88, 116–17, 133–34, 136, *169*, *174*, 189; Proust Questionnaire, *184*, 248–49; relationship with Sandra, 39, 78, 81–82, 84, *168*; sixtieth birthday celebrations, 257–58; stroke and other health issues, 134, 191–92, 250–53; on women in science, 141

—childhood: abuse and neglect from foster family, 7–8, 9–10, 14, 17; attempts to run away, 10; birth and early childhood, 3–4; departure from La Chaux-de-Fonds, 18; friendships, 10–11, 16–17; grape harvesting, 12–13; photographs, *161*, *163*; separation from mother and move to La Chaux-de-Fonds, 4–7; teenage years, 11, 14–18

—youth: employment in France, 28; employment in Germany, 20–22; French mandatory military service, 25–26, 27–28, 66; friendships, 24–25; loss of religious faith, 19, 20–21; move to Germany, 19; photograph, *163*; political views, 30, 45, 46; relationship with Ute, 29, 40; reunion with birth family, 26–27, 28–29; room in Kiel, *167*; US travels and American family, 37–40, *164*

—education: baccalaureate studies and exam, 21, 22–23, 29, 31, *165*; doctoral program, 71–72, 73–75, 76–78; Ghana research trip, 55–58; habilitation to direct research, 129–31, *171*; invention of mud-separating machine, 52–53; at Oceanographic

Institute (Kiel), 42–43, 44, 46–47, 48, 50–53, 59, 74; primary and high schools, 8–9, 13, 14–15, 18; Russian studies, 59; US research trip, 60–61; *Walther Herwig* research trip, 61–64
—career: 1980s pace and scope of work, 132–33, *171*; Africa and, 263, 264, 282; arrival and departure from ICLARM, 84–85, 86, 167, 188–89; awards and honors, 11, *177*, *183*, 249, 254–57, 262, 290, 298, 308; Chinese fisheries and, 240–43, 275–76; collaboration with Longhurst, 131–32, 139; current work, 308–9, 310–12; Cury and, *177*, 285–86; Darwin and, 243–46; data transparency and, 75–76, 124, 270–71, 274; on determining fish growth rates, 98, 100–101, 102–3; on determining natural fish mortality rates (M), 74–76; early employment struggles, 66; on ecolabeling, 293; Ecopath and, 139–40, 142, 143–44; environmental NGOs and, 212, 227; FishBase and, 151–53, 155, 158, 159; on fisheries as global disaster, 196–97, 198; on fisheries solutions, 313; on fishing down trophic levels, 206–10, 211–17; French fisheries and, 286–87, 288–89; in Ghana, 55–59; gill-oxygen limitation theory, 73, 76–78, 246, 298–300, 303; global fisheries study, 192–96; on growth of tropical fish, 132; Hilborn's rivalry with, 304–5, 307; Indonesian research and development work, 65–67, 68–71, *165*; on longline fishing, 231–32; on Malthusian overfishing, 94–96; on marine mammals, seabirds, and fisheries, 259–60; media and, 195–96, 208–10, 242, 248, 288–89; Peruvian collaborations, 123–26, 128–29; public engagement, 247–48; reconstruction of global catches, 260–62; research assistants and students, 249; San Miguel Bay and social dimensions of fisheries, 91–94; on sardinella stocks, 269–70; teaching at University of the Philippines, 101–2; on tropical fisheries, 84–85; at UBC, 189, 190–91, 212, 249; on West African overfishing, 271–72, 276–77

—works: overview, *182*, 308; *Darwin's Fishes*, 245–46; "Fishing Down Marine Food Webs," 176, 206–10, 211–17; *Gasping Fish and Panting Squid*, 298, 299–300; *Global Atlas of Marine Fisheries* (with Zeller), 262; *Mange tes méduses!* (Eat your jellyfish!; with Cury), 295–97; *In a Perfect Ocean* (with Maclean), 233, 235–36, 237–39; *On the Sex of Fish and the Gender of Scientists*, 304
Pauly, Gérard (brother), 26–27, 30, 88
Pauly, Gilbert (brother), 29
Pauly, Ilya (son), 78, 81, 87–88, 133, 134, *169*, *180*, 220–24, 311
Pauly, Jocelyne (sister-in-law), 26, 88
Pauly, Louis (stepfather), 6, 29, *175*, *181*, 233–34
Pauly, Renée (mother; née Clément), 1–2, 3–4, 6, 26–27, 37, *162*, *180*, 234
Peru: anchoveta fisheries management, 126–27, 128; anchoveta study with Pauly, 123–26; El Niño–Southern Oscillation (ENSO), 122; fish meal plants, 121; Instituto del Mar del Perú (IMARPE), 121, 128–29; oceanographic and fisheries research, 122–23
Petersen, C. G. Johannes, 98–100
Pew Charitable Trusts, 227–28, 229, 240, 249, 259, 309, 313
Pew Foundation, 276
Philippines: aquaculture, 103–4; Pauly's life in, 87–88, 116–17, 133–34, 136, *174*, 189; political tensions, 85–86, 89–90, 106, 108, 135–36; San Miguel Bay and social dimensions of fisheries, 90–94; Verde Island Passage, 112; Wade in, 86–87. *See also* International Center for Living Aquatic Resources Management
phytoplankton, 119, 192–94
Pimm, Stuart, 195
pirate fishing, 266
Pitcher, Tony, 189, 236, 301
Polovina, Jeffrey, 136, 137–39, 142
Pörtner, Hans-Otto, 299
positivism, 72
productivism, 264

Pullin, Roger, 86, 114–15, 116, 129, 187, 188, 189, 190, 252

Rabhi, Pierre, 297
racism, 19–20, 33, 57–58, 146, 189, 271
Ramón Margalef Prize in Ecology, 11
Ramsar Convention, 59
Randall, John, 140
reduction fisheries (fish meal plants), 121, 266, 267, 281, 294, 313
Reibisch, Johannes, 69
Reichert, Joshua, 227–28, 229, 233
Reimers, Dirk, 77
Reyher, Samuel, 43
Ricker, Bill, 122
Ricklefs, Robert, 84
Rockefeller Foundation, 83, 129, 313
Rognon, Marguerite, 10–11
Roqué, Mateo, 11
Rost, Dietmar, 198
Royal, Ségolène, 291
Rudolph, Wilma, 12, 13
Russell, Edward, 48
Russia. See Soviet Union

Sadovy de Mitcheson, Yvonne, 300
Sakumo Lagoon (Ghana), 56–57, 58–59
Samb, Birane, 269–70, 271
Samba, Alassane, 270, 280, 281
San Miguel Bay (Philippines), 90–94
sardinella, 57, 264, 267, 269, 270
Science (journal), 200, 202, 207–8, 213, 231, 243, 253–54, 276–77, 305
Scientific American (magazine), 254–55
Sea Around Us: Australian office, 312; beginnings of, 227–28; on Chinese fisheries, 240–43, 275; on climate change and species distribution ranges, 301–2; funding and staff, 249, 309–10, 312; on North Atlantic fisheries, 228–29; reconstruction of global catches, 260–62; reconstruction of West African catches, 277–78, 279–81
sea bass, 70
seabirds, 260, 266–67
SeaLifeBase, 158–59

Sea Shepherd, 199, 212, 280
Second World War, 1, 2, 3
Senegal: artisanal fisheries, 179, 264–65; catch reconstruction, 278; efforts against foreign fisheries, 279, 280; fisheries impacts on seabirds, 266–67; French fisheries researchers in, 263–64; global exploitation of fisheries, 265–66, 272; scope of overfishing, 267–68; Western exploitation of, 282–83. See also West Africa
Seto, Katy, 268
sharks, 231
Sharpless, Andy, 313–14
Shehadeh, Ziad, 86
shifting baseline syndrome, 197–98, 201–2, 205
Shomura, Richard, 136
Shul'man, Georgii Evgen'evich, 59
Sierra Leone, 268–69, 279
simple living, 297
slavery, 311n
Smith, Ian, 90, 92
South West Africa (Namibia), 23–24
Soviet Union, 59
Sparre, Per, 111, 139
Stakhanov, Aleksey, 241n
Starr, Steve, 36
statistics, Bayesian, 100–101
Stergiou, Konstantinos "Kostas," 158, 249–50, 308
Sumaila, Rashid, 236–37, 269, 309, 313
Système d'Information et d'Analyses des Pêches (SIAP), 271–72, 274, 276

tall poppy syndrome, 236
temperature, of oceans, 302
Thiao, Djiga, 267–68, 281–82
tilapia, 56–57, 60, 66, 294
Tomas, Tessie, 116
Torres, Armi, 155, 158
Torres, Francisco, 207, 212–13
trawlers, 61–62, 66, 92–93, 289–90, 290–91
Trends in Ecology and Evolution (TREE), 196–97
Trites, Andrew, 259

trophic levels: about, 142–43, 193; Ecopath and, 143; fishing down, 176, 206–10, 211–17, 231; marine trophic index, 215
tropical fisheries, 69, 84–85. See also coral ecosystems; Indonesia; Philippines
Trump, Donald, 32–33, 305, 310
Tsikliras, Athanassios, 158
Tsukayama, Isabel, 123–24, 126, 128, 170
tuna, 198, 265–66, 272, 279, 302, 314

Ulanowicz, Robert, 143, 144
uncontrolled fisheries, 266
United States of America: at 1960 Rome Olympics, 12; aid and development from, 83; armed forces segregation, 2–3; California fisheries, 121; Cornell University occupation, 35–36; education desegregation, 33–35; Pauly's travels in, 37–40, 60–61, 164
University of British Columbia (UBC): Institute for Oceans and Fisheries (Fisheries Centre), 228, 249, 309; Pauly and, 189, 190–91, 212
University of the Philippines, 89, 101–2, 154–55
upwelling zones, 119–21
Ursin, Erik, 139

Vakily, Michael, 91, 154, 268, 271
Vancouver (BC), 78–79, 189, 212
Venema, Siebren, 110–11, 134
Verde Island Passage (Philippines), 112
Vitousek, Peter, 192, 193
von Bertalanffy, Ludwig, 72–73, 76, 100

Wadden Sea, 203
Wade-Pauly, Sandra (Pauly's wife): background, 79–81; Daniel's stroke and other health issues, 250–52, 253, 308, 309; family life and parenting, 88, 133, 169, 218; gill-oxygen limitation theory and, 299; International Cosmos Prize and, 256, 257; in Philippines, 86–87; relationship with Daniel, 39, 78, 81–82, 84, 168, 310; in Vancouver, 219
Wallace, Alfred Russel, 69

Walsh, John, 122
Walters, Carl, 147–48, 232, 257, 304, 305
Walther Herwig (research vessel), 61–64
Wanless, Ross, 267
Watson, Reg, 216, 229, 240–41, 260, 301, 302, 305
West Africa: about, 263; Canary Current system, 264; catch reconstruction and controversy, 277–78, 280–81; Chinese fisheries in, 266–67, 275–76; data sharing and, 273–74; fisheries impacts on seabirds, 266–67; global exploitation of fisheries, 265–66, 272; overfishing concerns, 281–82; pro-fishing lobby, 276–77; sardinella fish stocks study, 269–70; Sea Around Us impacts and efforts against foreign fisheries, 279–80; SIAP project on overfishing, 271–72. See also Ghana; Senegal; Sierra Leone
Western Sahara, 279
whaling, 276
Whitfield, Ed (Pauly's cousin), 32, 33–37, 38, 166
Whitfield, Robert (Pauly's cousin), 37, 60
Whitfield, Winifred (Pauly's aunt), 33, 37, 38, 164
Winberg, Georgii G., 59, 73
Woller, Gerd, 75
women, in science, 141
WorldFish Center. See International Center for Living Aquatic Resources Management
World Wildlife Fund (WWF), 276, 293
Worm, Boris, 204, 229, 230–32, 253–54, 262, 305
Wosnitza, Claudia, 125

Zeller, Dirk, 260, 262, 305, 311–12

DAVID SUZUKI INSTITUTE

THE DAVID SUZUKI INSTITUTE is a non-profit organization founded in 2010 to stimulate debate and action on environmental issues. The Institute and the David Suzuki Foundation both work to advance awareness of environmental issues important to all Canadians.

We invite you to support the activities of the Institute. For more information please contact us at:

David Suzuki Institute
219 – 2211 West 4th Avenue
Vancouver, BC, Canada V6K 4S2
info@davidsuzukiinstitute.org
604-742-2899
www.davidsuzukiinstitute.org

Cheques can be made payable to The David Suzuki Institute.